Book
about
BEES

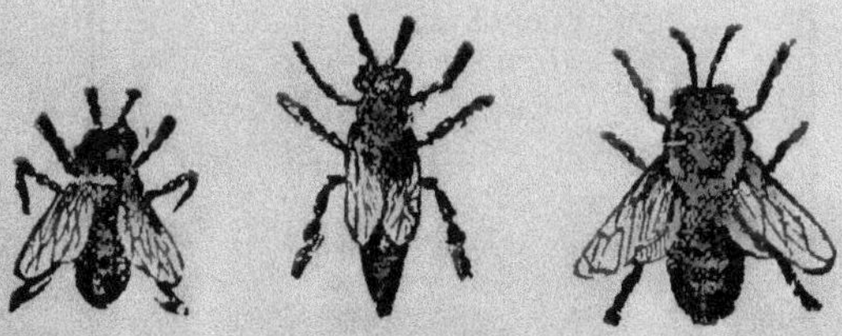

SF523
J54

ALBERT R. MANN
LIBRARY

NEW YORK STATE COLLEGES
OF
AGRICULTURE AND HOME ECONOMICS

AT

CORNELL UNIVERSITY

EVERETT FRANKLIN PHILLIPS

BEEKEEPING LIBRARY

The original of this book is in
the Cornell University Library.

There are no known copyright restrictions in
the United States on the use of the text.

http://www.archive.org/details/cu31924003691544

A BOOK ABOUT BEES.

The Cottage Bee-keeper's Home. [*From a Drawing by Charles Jenyns.*]

A Book about Bees.

Their History, Habits, and Instincts;

TOGETHER WITH

The First Principles of Modern Bee-keeping

FOR YOUNG READERS.

BY

REV. F. G. JENYNS,

Rector of Knebworth; Member of the Committee of the
British Bee-keepers Association.

With Introduction by the BARONESS BURDETT-COUTTS.

*Published at the Request, and under the Sanction, of the British
Bee-keepers' Association.*

LONDON:
WELLS GARDNER, DARTON, & CO.
2 PATERNOSTER BUILDINGS.
1886.

@ 39528

Gratefully Dedicated

TO THE

BARONESS BURDETT-COUTTS,
President of the British Bee-keepers' Association.

WHOSE NAME IS A HOUSEHOLD WORD THROUGHOUT THE LAND
IN CONNECTION WITH ALL GOOD WORKS;
AND WHO, WHILE DEVOTING HER WEALTH AND ENERGIES,
WITH THE MOST UNBOUNDED CHARITY, IN THE WIDEST FIELDS OF
BENEVOLENCE,—
IS YET ALWAYS READY TO JOIN HEARTILY IN EVEN THE
HUMBLEST EFFORTS DIRECTED TO THE BENEFIT AND ADVANCEMENT
OF THE INDUSTRIAL CLASSES:

AND,

WITH THIS END IN VIEW, HAS EVER SHEWN
THE WARMEST SYMPATHY WITH THOSE WHO SEEK TO PROMOTE
THE MORE EXTENSIVE KNOWLEDGE OF BEES,

AND

THE PRACTICE OF INTELLIGENT BEE-KEEPING.

INTRODUCTION.

STRATTON STREET,
December 6th, 1885.

DEAR MR. JENYNS,

I am much gratified by your kind wish to dedicate to me your valuable educational contribution on Bees and Bee-keeping.

This industry has made a rapid progress under the fostering care of the British Bee-keepers' Association, and the untiring zeal of its late esteemed Honorary Secretary, the Rev. H. R. Peel. Bees now rank as fellow-workers in the objects of the Royal Agricultural Society; and, through the formation of the British Honey Company, they are linked with those industrial and commercial projects, which seek to promote the food supply of the people at large, and to render it plentiful and wholesome.

In this book you point out very justly that Bees and Bee-keeping can be made subservient to an educational purpose, and also possess an interest under this aspect of no small value.

Religious instruction, with certain other definite subjects, such as Reading, Writing, Arithmetic, History, and Geography, must form the foundation of all teaching in schools, but whilst these must be equally taught in all, there are always specific subjects to be added; and it will probably be found advisable to adapt these, more frequently than is done at present, to the circumstances of the locality in which the schools are situated; so that information should be given in agricultural, manufacturing, and inland districts, somewhat differing to that given at the sea-board and in towns.

Your lessons on Bees are admirably calculated to point out how such information can be given, without adding any additional pressure to the already high-pressure system of education in vogue at the present time; and to direct attention to the means of conveying instruction to children in matters which would naturally enter largely into their occupations on leaving school.

I trust your interesting little book may be a pioneer in this direction, and give an impulse to Reading lessons calculated to give to children information of an accurate and interesting kind, bearing, in some measure, on their daily life, and strengthening their powers of observation on things familiar to their eyes and hands, yet with which they are, through lack of observation, imperfectly acquainted.

INTRODUCTION.

May your Bee lessons have yet a wider mission! May they promote Manuals which will lead the mind to the Creator, whether they treat of His wonders in nature, or of those wonders in art and in those handicrafts, which He has given to man the marvellous power to exercise! May our children by such means be led to appreciate the order and variety impressed throughout His Creation; and so not only learn to labour usefully, but to derive that peaceable pleasure which instruction such as this affords, to sweeten and lighten the occupations of daily life.

In conclusion, wishing all success to your effort, I would end with the sweet words of a shrewd observer, as well as a single-hearted Christian, and say to the Children 'Familie' who may con your Bee lessons,—

> 'First Peace and Silence all disputes controll,
> Then Order plaies the soul;
> And giving all things their set forms and houres,
> Makes of wilde woods sweet walks and bowres.
>   HERBERT.

I am,

Yours sincerely,

BURDETT-COUTTS.

AUTHOR'S PREFACE.

IN writing the following pages I have had no wish whatever to add another to the many existing 'Guides' to the management of Bees; neither have I attempted in any way to produce a scientific treatise. I have simply endeavoured to write an introduction to the subject suitable to young readers; and, while impressing the importance of habits of observation, have sought to unfold to them one little page of the vast Book of Nature; and, by showing some of the simplest of the many wonders of bee-life, to give them that interest in the subject which may lead them to desire to know more, and, afterwards, to take up Bee-keeping for themselves, with that knowledge which, while it adds tenfold to the interest, is more or less absolutely necessary to make it profitable.

But, while the book is thus in great measure introductory, and is primarily intended for the young, it is hoped that it may not be altogether uninteresting to those of riper years, and may furnish them with some inducement to proceed to the investigation of the science of the subject, and to Bee-keeping in its most modern and advanced methods.

AUTHOR'S PREFACE.

Its preparation was undertaken at the request of the Committee of the British Bee-keepers' Association, who felt that, in their efforts to promote intelligent Bee-keeping as a national industry, the young should not be neglected, and that there ought to be a book suitable for use—where found practicable—as a Reading-book in Schools; or, at all events, one likely to find its way into the hands of those young people of all classes, who soon will be old enough to become bee-keepers.

I have with pleasure, and gratefully, to record the assistance I have received in its preparation from my friends the Rev. G. Raynor and the Rev. J. Lingen Seager—the well-known bee-keepers—who have most kindly supervised my work. I am also much indebted to Mr. Cowan, Chairman of the British Bee-keepers' Association, for much kindness and very valuable advice. I have also to thank the Association, Mr. Cowan, Sir J. Lubbock, Mr. Neighbour, Mr. Baldwin, and Mr. Walton, for the use of illustrations belonging to them. My especial thanks are also due to Mr. W. B. Carr and Mr. Charles Jenyns for many original illustrations.

F. G. J.

Knebworth Rectory, Dec. 1885.

LIST OF ILLUSTRATIONS.

	PAGE
THE COTTAGE BEE-KEEPERS' HOME	*Frontispiece*
WILD BEES AND FLOWERS	23
ITALIAN BEE	26
QUEEN, WORKER, AND DRONE BEES	36
THE QUEEN AND HER ATTENDANTS	38
QUEEN CELLS IN DIFFERENT STAGES	43
THE COTTAGER AND HIS BEES	46
HIVING A SWARM	51
THE HEAD, THORAX, ABDOMEN, OF A BEE	61
WING OF A BEE	63
EGG AND LARVÆ OF THE BEE	69
LEG AND POLLEN-BASKET. *Two Illustrations*	77
TONGUE. *Two Illustrations*	78
WING, SHOWING HOOKLETS	80
BEE AND ITS STING	82
STING HIGHLY MAGNIFIED	83
HEAD AND ANTENNÆ	84
COMB, WORKER AND DRONE	91
COMB FOUNDATION	91

LIST OF ILLUSTRATIONS.

	PAGE
Possible forms of Cells — Circular, Square, Triangular	92, 93
Economy of the Hexagonal Form	94
Leg with notch, Magnified	108
The Cottage Hive of Helpful Children	113
Observatory Hive	120
Straw Hive and Super	120
Frame	124
Hive showing Frames in Position	135
Frame, Empty, and with Foundation	136
Frame filled with Comb	137
Section of a Hive with Frames	138
Queen Cages	143
Bee Veil	145
Smoker in Use	145
Sections with Foundation	148
Sections in Rack	148
Hive with Sections in Position	149
Extractor. *Two Illustrations*	151
Driving Bees from a Skep	159
Flowers — Cherry, Buttercup, Apple, Meadow Geranium, Wood Sage	178—181
Bees and Orchis	183
Bee Tent	199

CONTENTS.

CHAPTER I.
HABITS OF OBSERVATION 1
    Wonders of Nature around us Everywhere—The Scotch Naturalist—Professor Henslow—The Microscope—'Eyes and no Eyes.'

CHAPTER II.
BEES TO BE OBSERVED AND TREATED WITH GENTLENESS 7
    Bees: A Subject of Interest—When and Why they Sting—To be treated with Kindness—Coleridge.

CHAPTER III.
THE WORK OF THE BEE—INTRODUCTORY . . . 10
    'Mind your own Business'—The Bee works itself to Death—Sir J. Lubbock's Observations on the Work of a Bee.

CHAPTER IV.
THE BEE'S BUSY LIFE—CONTINUED 13
    The Bee does not 'rust out:' contented, patient, persevering—The Donkey at Carisbrooke Castle.

CHAPTER V.
COMMUNITY OF BEES IN A HIVE 15
    Rooks and other Birds congregate—Bees in the Hive as a Community—Ants—Sir J. Lubbock—Beavers—The 'Song of the Bees.'

CHAPTER VI.

DIFFERENT KINDS OF BEES 21

Vast Numbers of different kinds of Bees—Humble Bees: their great Use; taken to New Zealand—Solitary Bees: the Leaf-cutting—The Mason.

CHAPTER VII.

VARIETIES OF THE HONEY BEE 25

Some Honey Bees more valuable than others—Italian Bees—Cyprian and other Bees.

CHAPTER VIII.

AMERICAN BEES 27

The Wild Bees of America—Bees and Monkeys—Bee-hunting in America—The Bee-line—Instinct—Story of a Cat.

CHAPTER IX.

BEES IN THE OLDEN TIME 30

Bees and Honey, mention of, in the Bible — Virgil, and others—Huber, and his Life.

CHAPTER X.

THE INHABITANTS OF THE HIVE—INTRODUCTORY . 34

Modern Discoveries—The different Kinds of Bees in a Hive—The Workers, and their Number — The Queen as the Mother of the Colony—Old Errors regarding the Queen—The Drones, noisy and idle.

CHAPTER XI.

HOME OF THE HONEY BEE—INTRODUCTORY . . . 41

The Brood Nest and its Contents—The Queen-cell—The Store-room of the Hive—Every Portion of Space made use of—Everything in its place.

CHAPTER XII.

GENERAL OUTLINE OF THE HISTORY OF A HIVE . . 47

A Swarm issuing — The Swarm secured and hived — Its Work during the Summer — Its Condition in Autumn and Winter.

CONTENTS.

CHAPTER XIII.

A TALE OF DESTRUCTION 53

The Goose and its golden Eggs—The old Custom of destroying the Bees—An inhuman and foolish System—Thomson's 'Seasons.'

CHAPTER XIV.

THE NATURAL HISTORY OF THE BEE—INTRODUCTORY . 56

Some Knowledge of the Bee's Natural History necessary—Observe everything, and have a Reason for all you do—Illustrative Examples of successful Observers and Discoverers—Franklin, and others—Hervey and the Circulation of the Blood.

CHAPTER XV.

THE NATURAL HISTORY OF THE BEE—CONTINUED . 59

The Classification of Animals—The Bee's Position in the Insect World—Recapitulation, and Diagram.

CHAPTER XVI.

THE TRANSFORMATION OF INSECTS 66

The Egg, Larva, Pupa, Imago—The Provision of Nature for Food for the Larvæ of different Insects—Examples.

CHAPTER XVII.

THE NATURAL HISTORY OF THE BEE—CONTINUED . 69

The Process of Transformation from the Egg to the perfect Bee in the cases respectively of Worker, Drone, and Queen—'Royal Jelly' and its Effects on the Larva—The difference between the Queen and Worker illustrated by that between the Greyhound and Pug.

CHAPTER XVIII.

STRUCTURE OE THE BEE ADAPTED TO ITS WANTS . . 74

Examples of the same Adaptation to be found in Animals, Birds, Fishes, and Man himself—The Leg of the Bee—The Tongue—The Wings.

CHAPTER XIX.

THE SAME SUBJECT—CONTINUED 81
    The Sting of the Bee—The Sting under the Microscope—The Antennæ—The Antennæ a means of Communication.

CHAPTER XX.

THE SAME SUBJECT—CONTINUED 86
    The Differences in Formation of Queen, Worker, and Drone—A Worker sometimes lays Eggs.

CHAPTER XXI.

COMBS, AND THE FORM OF CELLS 88
    Examples of Engineering Skill—The Perfection of the Hexagonal Shape of Cell—Mathematics of the Hive—Combs must have great Strength—How this Strength is obtained with the least material possible—How Worker and Drone Cells are joined together.

CHAPTER XXII.

MORE ABOUT WHAT THE BEES DO 97
    The Queen, her Work, and Instincts—The Number of Eggs she lays—The Drones and their Use—Drones destroyed.

CHAPTER XXIII.

THE SAME SUBJECT—CONTINUED 102
    The Worker's Work—Shakspeare on the Honey Bee—How far Bees go from the Hive—The Rapidity of the Bee's Flight—Bees in any one Journey keep to one kind of Flower.

CHAPTER XXIV.

MORE ABOUT WHAT THE BEES GATHER — HONEY, POLLEN, AND PROPOLIS 106
    Honey derived from the Nectar of Flowers — How the Bee removes Pollen from its Tongue — Propolis, whence gathered—Curious Instances of its Use—A Snail destroyed—A Slug buried.

CHAPTER XXV

WAX, AND HOW THE BEES MAKE IT. 110

Wax produced from Honey—The Time and Labour necessary for the Process—All Workers, except the young Bees, are Wax-makers—The Work of young Bees—The Children of the Hive most Useful.

CHAPTER XXVI.

NIGHT-WORK AND VENTILATION 114

Bees never Sleep—The Necessity of Ventilation, and how the Bees provide for it—The Guard at the Gate.

CHAPTER XXVII.

THE DIVISION OF LABOUR IN THE HIVE . . . 116

Every Bee has its Work—Their Attention to little Things—The Importance of little Things—Other Examples of the Wonders done by little Creatures—The Hive a Savings Bank.

CHAPTER XXVIII.

MORE ABOUT THE OBSERVATION OF BEES . . . 120

The Observatory Hive—Huber's 'Leaf' Hive—Sir J. Lubbock's Observations of Bees and their Sense of Colour.

CHAPTER XXIX.

INTRODUCTION TO MODERN BEE-KEEPING . . . 123

Boys and Girls may keep Bees—Best to begin in a small Way—The Necessity of Common Sense and Perseverance—Whatever you do, do it thoroughly—Good Management always pays—Virgil's Story of the old Gardener.

CHAPTER XXX.

FIRST PRINCIPLES OF BEE-KEEPING 127

This book not 'a Guide-book'—The Golden Rule for successful Bee-keeping—The best kind of Straw Hive and how to begin with it—The Disadvantages of the Straw Hive.

CONTENTS.

CHAPTER XXXI.

PAGE

THE FRAME-HIVE AND THE PRINCIPLES OF ITS CONSTRUCTION 133

The Frames and their Use—Comb Foundation, how it is made and its Use—What is essential in a good Hive—Homemade Hives—Examples of Ingenuity.

CHAPTER XXXII.

ADVANTAGES OF THE FRAME-HIVE 141

Interchange of Frames—Introduction of young Queens—Extraction of Honey—Increase of Colonies—Enemies destroyed—Something about Bee-veils and Smokers.

CHAPTER XXXIII.

SUPER HONEY AND THE EXTRACTOR 146

Bees will swarm unless they have Plenty of Room—Sections and Supers how prepared and placed on the Hive—Supers filled and refilled—The Extractor, its Construction, Use, and Principle.

CHAPTER XXXIV.

MORE ABOUT SWARMS AND CASTS 152

'A Swarm in May worth a load of Hay'—Second Swarms or Casts—When and how brought about, and how controlled—The Combat of Queens—Note by the Rev. George Raynor.

CHAPTER XXXV.

THE BUSY BEE-KEEPER IN SUMMER, AUTUMN, AND WINTER 156

The Bee-keeper in Summer—The Quantity of Honey to be obtained from well-managed Hives—Note by Thomas W. Cowan, Esq.—The Bee-keeper in Autumn—The Operation of 'Driving' and its Use—The Bee-keeper at rest in Winter.

CHAPTER XXXVI.

THE CONNEXION BETWEEN FOOD AND WARMTH . . 162

Food as the Producer of Flesh and Warmth—The Hibernation of Animals—The Bee does not hibernate—Warmth in the Hive sustained by many Bees and Plenty of Food—The Necessity of pure Air.

CHAPTER XXXVII.

THE BUSY BEE-KEEPER IN SPRING 165

Spring-time and the 'Song of the Bees'—The Importance of early Brood—Stimulative Feeding; to be done with great Judgment—Artificial Pollen and its Use—Spreading Brood and its Dangers—The Importance of 'Brains' in Bee-keeping.

CHAPTER XXXVIII.

DISEASES AND ENEMIES OF BEES 169

'Prevention better than Cure'—The Importance of Attention to the Rules of Health—'Foul Brood' the most fatal of Diseases—The Wax-moth and other Enemies—Robber Bees.

CHAPTER XXXIX.

THE USES OF HONEY, WAX, AND PROPOLIS . . . 172

Honey as Food—Used formerly instead of Sugar—Mead and other Drinks—Variety of Food necessary for our Bodies—Nourishment in Honey—Wax and Propolis put to many Uses—Bees used to quell an angry Mob.

CHAPTER XL.

FLOWERS IN RELATION TO BEES 176

Bees as useful to Flowers as Flowers to Bees—Some parts of a Flower described—What, in a general way. is necessary for Fertilisation—Provisions of Nature to secure cross Fertilisation—Without it Plants die out or degenerate—

FLOWERS IN RELATION TO BEES—*Continued.*

Bees the Handmaidens of Nature to the end in view—The Use of Honey to the Flowers as an Attraction to Bees and other Insects—The Colours of Flowers attract Insects—Flowers fertilised by the Wind are Colourless—The subject full of Lessons of Divine Truth.

CHAPTER XLI.

THE IMPORTANCE OF BEE-KEEPING 186

Except in a few Localities almost any Number of Bees may be kept with Profit—Bee-keeping abroad and in America practised on an extensive Scale—A great Waste in England of Nature's Gifts—Honey everywhere: Bees required to gather it—Considerable Profit available for Cottagers—Bees required for the profitable Culture of Orchard Trees, and for Garden and Field Crops—'Welcome to the Bee.'

CHAPTER XLII.

SUPERSTITIONS WITH REGARD TO BEES 191

How Superstitions become prevalent—Bees must be bought with Gold—A Cornish Superstition—'Tanging:' its possible use in former Times—Bees in Mourning—Story by Rev. G. Raynor.

CHAPTER XLIII.

BEE-KEEPERS' ASSOCIATIONS AND SHOWS . . . 195

Bee-keeping Literature—Associations: their Object and Work—A County Show—The Gathering of Exhibitors—Success or Failure—The Bee-tent—Conclusion.

A BOOK ABOUT BEES.
FOR YOUNG READERS.

CHAPTER I.

HABITS OF OBSERVATION.

IF we travel through England we find most varied scenery; some of it beautiful with mountains, valleys, woods, and water; and some of it flat, bare, and wild. But, whatever the character of the country, we may always find in the works of Nature much that is indeed very beautiful and wonderful, and much to make us full of good thoughts. We learn of the great Creator by all that we see of His works and creatures.

> 'There is a Book, who runs may read,
>   Which heavenly truth imparts,
> And all the love its scholars need,
>   Pure eyes and Christian hearts.
>
> 'The works of God above, below,
>   Within us and around,
> Are pages in that Book, to show
>   How God Himself is found.
>
> 'Thou, who hast given me eyes to see
>   And love this sight so fair,
> Give me a heart to find out Thee,
>   And read Thee everywhere.'

Probably you have been to London, and doubtless you were astonished when first you saw its great sights. What did you like best? Perhaps you can hardly tell, for you saw so many things to interest you. You saw its long streets with the shops, and crowds of busy people; and you saw its grand buildings—Westminster Abbey, and St. Paul's Cathedral, and the Houses of Parliament, and the Palace of the Queen; and you saw the river Thames, and the bridges, and the great ships. And then, perhaps, you went to the Zoological Gardens and saw the elephants, and lions, and tigers, and the monkeys, and the birds; and when you returned home you thought you had never before seen anything so wonderful.

Yes; but every day, and all around you in the country, are many things to be seen quite as beautiful and wonderful, if only you will open your eyes to look for them, and take trouble to learn the nature, history, and use of what you see.

There is a story of one who is called 'the Scotch Naturalist,'—but who was only a poor and very needy shoemaker,—who loved all he saw in Nature so much that, after a long and hard day's work at his trade, he used to go out for long walks of many miles into the fields and moors, and by the rivers, and stay out all night, lying perhaps in cold and wet, on purpose to observe the habits of some little animals, or to collect specimens of plants and insects to take home and preserve in the wonderful collection which he made. By night, as well as by day, he saw wonders in many things which other people thought little of, and great and fresh wonders continually.

You cannot indeed go out at night as he did; but, for instance, in any walk or excursion in the country you can gather a little flower; and if you only knew how to pick that flower to pieces (and you may soon learn), and were taught the uses of its several parts, and how all fit together and grow together, and are necessary one to another, and provide the seed which grows into other plants of the same kind another year, you would indeed be surprised and interested.

> 'Nature, enchanting Nature, in whose form
> And lineaments divine I trace a hand
> That errs not, and find raptures still renew'd,
> Is free to all men—universal prize.
> Strange that so fair a creature should yet want
> Admirers, and be destin'd to divide
> With meaner objects ev'n the few she finds!'
> COWPER.

Some few years ago, at the village school in the parish of Hitcham, of which Professor Henslow, the well-known botanist, was the Rector, the elder children used once a-week to bring flowers and other things to school, and were taught to examine and preserve them; then, at the end of the year, there used to be a show of all they had found and prepared, and prizes were given; and thus the children's eyes became very sharp to search for and find little things that perhaps you would not think worth looking at.

And here I would say that if you want to see the wonders of Nature, you must always remember that many of the greatest wonders are found in the smallest things. You must not think that a thing is wonderful simply because it is large. When you have seen an elephant you have probably been astonished at the size of its massive legs; but really the leg of a little fly is quite as wonderful—just as marvellously made, and just as beautifully fitted for what it has to do.

Very many, however, of these wonders of Nature are so minute that it is quite impossible to see them without the aid of a microscope. This, as probably you know, is a very beautiful instrument with several glasses, made and fixed in a particular way, so that when you look through them at any very small object, such as a fine hair, it is so magnified that it looks almost as large as a walking-stick; or it will make the very small tongue of a bee appear as a long thing with many joints; or it will show you the sting of a bee as large as and much more finely pointed than any needle.* Again, if you take a drop of water out of some ditch and put it under the microscope, you will see it full of little animals, like very odd-shaped fishes, swimming about and perhaps eating one another, although without the microscope the drop of water may appear quite clear and to have nothing in it at all.

But all the same, do not think you must have a microscope to see a great many of the things of which I have been speaking. Only use your eyes as you walk about; and when you see anything that attracts your attention, try and find out, and answer the questions, 'What is this?' and 'Why is this?' and 'What is its use?' You may always be sure that everything in Nature has some use and serves

* Illustrations of the tongue and sting of the bee will be seen at pages 78, 82, and 83.

some wise, although often mysterious purpose of the Divine Maker.

As examples of what I mean, here are two very common things for you to explain if you can. You have often seen a fly walking on the ceiling of the room, but perhaps you have never thought how it can do this with its head and body downwards. You could not do it, neither could the cat; but the little fly does it easily. Now, how is this? Perhaps you know; but if not, you must try and find out. Ask your teacher, or read some book, and you will learn how very wonderfully its little legs are made, with a vast number of 'sucker hairs' clothing the pad of each foot, exactly fitting it for what it has to do.

Again, as you walk along some country road you pick up a little round stone, quite smooth, without any sharp edges. Now, why is it smooth? how came it so? Can you tell? Do you know that once upon a time it was in the sea and was rubbed about by the great waves, one stone against another, till it became quite smooth? Ask the stone to tell you its wonderful story. I am sure you will like to hear it.

Even a drop of water could tell you a marvellous tale of wonderful journeys and changes. Yes, listen thus to Nature's voice, and, as the great poet, Shakspeare, says, you will find—

'. . . . Tongues in trees, books in the running brooks,
Sermons in stones, and good in everything.'

Probably you know the old story called 'Eyes and No Eyes'—about two boys who went out one day to spend a holiday, and, without knowing it, went the same way, across the same fields, over the same moor, by the same stream, near the same sand-pit, and up the same hill.

When they came home at evening, they were asked what they had seen. Dick, who came home first, said he had had a very dull, stupid walk, and had seen nothing of any consequence. Will, on the contrary, had seen, he said, so many odd and wonderful things that he had never more thoroughly enjoyed himself. He had seen the woodpecker at work, and the lapwing feigning lameness to draw him from her nest, and many other birds, and a snake, and some beautiful flowers, and had found some curious fossils, and had brought home his handkerchief quite full. And yet Dick had found nothing even to look at! The fact was, the one had walked about with his eyes shut, and the other had kept his wide open.

And, however long we live in any place, it is the same—there is always much that is fresh to see. White, the naturalist of Selborne, says that 'that district produces the greatest variety which is the most examined;' and another naturalist observes, that 'so rich is nature that a man born a thousand years hence will still find enough left for him to do and notice.' But 'many waste a whole life without ever being once well awake in it, passing through the world like a heedless traveller, without making any reflections or observations, without any design or purpose beseeming a man.'

CHAPTER II.

BEES TO BE OBSERVED AND TREATED WITH GENTLENESS.

Now this book is about bees, their homes and habits; and very curious and wonderful all these things are—

> 'A picture wonderful, an insect race,
> Their customs, manners, nations, I describe.'
> Virgil.

But we shall not see much of these wonders unless we keep our eyes open. Bees, indeed, must be observed, closely watched, read about, and thought of, before they can be understood. And the more we do this, the more wonderful and interesting we find them in all their ways and doings.

To a certain extent, of course, they are familiar to us all. It is a pretty sight we think, to see the hives in a cottage garden in a snug corner, surrounded with sweet-smelling flowers, and, on sunny days, to hear the hum of the bees as we see them flying in and out of their homes, or as we see them darting from flower to flower—the 'busy bee.'

But to many people they are of no further interest. Perhaps, indeed, they only think of bees as gathering honey, for which they do not care, or as having sharp stings, of which they are afraid.

I hope, however, you will not thus think of them. I only wish you could come with me, and look at the inside of a hive, and see what it is like, and what goes on there; and I am sure you would be interested.

As, however, you cannot do this, I must try and describe some of these things. And then, after a time I have no doubt, some friend will show you a hive, and its bees, or you will be able to look for yourself —take a hive full of bees in your hands, and thoroughly examine all that is inside, and touch the bees, and let them crawl over you; and all this without getting a single sting.

This may sound wonderful, but it is not really so. In fact, there is no difficulty about it; only, first of all, you must understand something of their habits, and then, of course, take some precautions. Many people wear a veil, and it is well to do so; but that which is of chief importance is quietness—that we go to the hive gently, and without noise or bustle, and take great care not to jar the hives, or to breathe upon the bees. Carefully attending to these things, and then using, as you will be told at a future time, a little smoke, the bees will allow us to do almost anything we like.

The best way, in short, is to treat them in some measure as pets; and even children may thus keep bees. I dare say some of you have pet animals at home—perhaps a kitten, or it may be a canary, or a goldfinch; and, if you deserve to have it, you treat it kindly, you feed it with the right food, you speak to it coaxingly, you guard it from its enemies. Well, in the same way almost, you may make pets of bees. They will not, indeed, come to you when you call, but they will, by some means, know that you are their friend, and will treat you as such. If you are kind to them, they will be kind to you.

In fact, bees seldom sting except in self-defence, or in defence of their homes. If you see one on a flower hard at work, it will never fly at you. Bees thus occupied never think of stinging, unless they are touched. And, even at their hives, they will but seldom attack us, unless by our own fault we make them angry, and they think their home and young are in danger. And can we find fault with them for this? Is it not most natural? Is it not praiseworthy?

I do not say, however, that bees will never sting without just cause, for I must confess that some are very easily provoked. There are cross bees just as there are bad-tempered people, very soon put out, and very resentful. Generally speaking, however, the temper of our bees mainly depends upon our own good or bad management.

Always remember then, that just as you get your companions to be good friends to you, or as you can get some pet animal to do almost anything you wish, if only you act with kindness, so bees also, must be treated in the same kind, gentle way, and then you will all be good friends together. Treat them, in short, as if you loved them; treat them as God's creatures.

> 'He prayeth well, who loveth well
> Both man and bird and beast.
> He prayeth best, who loveth best
> All things both great and small;
> For the dear God who loveth us,
> He made and loveth all.'
> COLERIDGE.

CHAPTER III.

THE WORK OF THE BEE—INTRODUCTORY.

WELL, we are to keep our eyes open. This was our first lesson; and now we have heard how kindness and gentleness will help us with bees, as with every other creature of God. And these are two very good things to learn; but the bees have many other lessons to teach us, and before we go on to speak of other things, we will listen to their pleasant hum, and see if we can make out anything they have to tell us.

I told you just now that a bee flying from flower to flower will never sting you if you leave it alone. Only if you hinder it, tease it, touch it roughly, will it at last get angry. Yes, and so what I think it says in its hum is this: 'Do, pray, mind your own work, and let me attend to mine. I don't want in the least to interfere with you, and only wish myself to be let alone. I have much to do. Pray do not stop me.' And this is a capital lesson, for we all have our work to do; and whether it is a great or little thing, the way to do it well is to stick to it, and to give it our whole attention. I dare say you often have lessons at school, or things to do at home that you find hard or troublesome. Well, follow the example of the bee, and, while you do not interfere with others in their work, don't let them stop you. Always remember that your task or duty, whatever it is, is of the first importance.

Then I hear something else in the hum of the bee as it passes me so swiftly on its way from the hive to the flower and back again; I fancy I hear it say, 'I am very busy, but at the same time I am very happy.'

It certainly is a very busy bee. Let me give you some idea of what its work is, how busy its little life. In the first place, it works so hard that it does not live long. In spring and summer-time, when there is much honey to be had, and a great deal of work to do, its life is a very short one, perhaps not more than from six to eight weeks. And to show you how this shortness of life is caused by hard work, you must know that a bee born in autumn, at the beginning of September, will live all through the winter, and generally during the first months of spring, that is to say, from six to eight months—as many months, you see, during its winter rest, as it lives weeks in summer, when it is hard at work.

The daily work of a bee in summer is something most remarkable. Go into the garden, when the sun rises early, and you will very soon see the bee come out and begin its day. And when it has found a place—perhaps it found this the day before—where honey, or whatever it wants, is to be had, whether it is far off or near, it begins to go backwards and forwards to the place as fast as it can fly. It has been found out that if the place, where it can get its food, is tolerably near, it will go backwards and forwards as many as eight or ten times in an hour.

Sir John Lubbock, of whose observations you will hear more, has made many experiments about this, and the way he contrived to find out the truth was as follows: In the first place, he got a few of his bees to come to some honey which he put ready for them at no great distance from their hive. He then marked one of the bees with a small spot of red paint, and another with perhaps a spot of blue; and thus knew his little friends again quite well when they came back for some more of his sweets.

Then he watched them carefully, putting down on paper the exact time when, for instance, his little red-painted friend came to the honey, and how long it was there feeding itself, or gathering its store to take home, and again noting when it flew away, and when it came back. So he watched it all through the day, and for days together, and thus knew at last how many journeys and visits to the honey his little bee made in the course of the day.

Then he did the same with other bees, and so at last by this means could pretty well guess what is generally the daily work of a bee. Sometimes, for one cause or another, his bee did not make so many journeys as at other times, but, generally speaking, its day's work was something as follows:—It would come to the honey very early, at six o'clock, or earlier according to the weather. It would then stay at the honey about two minutes loading itself, and then, flying away, would be gone about six minutes, in which time it went home, unloaded what it carried, and made its way back. Then again, it immediately began to load itself once more, taking about the same time, and going off again as at first.

This would go on hour after hour, so that perhaps

it would make nearly one hundred journeys in the day. Is not this a wonderful story of hard and persevering work? And when, at night-time, or on bad, rainy days, the bee was at home, we must not think it was idle. You will hear at a future time of what the bee does at home and at night; but now I only want you to think of the busy bee as you see it flying backwards and forwards, that so you may know something of what it does, and how hard it works. and the reason why, as I told you, the bee's life is not a very long one.

CHAPTER IV.

THE BEE'S BUSY LIFE—CONTINUED.

THERE is an old saying that 'it is better to wear out than to rust out,' which means that anything is better than an idle life. A thing that wears out, wears out by work, by being used, by fulfilling the purpose for which it was made, as, for instance, a spade, or a plough, or a knife. These wear out after a time by being constantly used, and it is far better that they should thus wear out, than be laid by, and so at last get rusty and useless. They have done their work when worn out, they have been of no use at all if they have only rusted out.

I think the bees must know something of this old saying. Most truly they do not 'rust out,' but 'wear out;' and if we are at all like the bees our lives will not be lives of idleness. We shall not get rusty for want of work.

> ' I come at morn, when dewdrops bright
>   Are twinkling on the grasses,
> And woo the balmy breeze in flight
>   That o'er the heather passes.
>
> ' Deem not these little eyes are dim
>   To every sense of duty ;
> We owe a certain debt to Him
>   Who clad this earth in beauty.
>
> ' And, therefore, I am never sad,
>   A burden homeward bringing,
> But help to make the summer glad
>   In my own way of singing.
>
> ' And thus my little life is fixed
>   Till tranquilly it closes.'
>
> CHAMBERS' *Journal*.

Indeed, as they work unceasingly day after day, doing the same thing, I do not think we ever can hear them say, in bee language, whatever it is, ' Oh, I am tired of all this ! It is just the same thing every day ! It is so dull to do it again and again !' The cheerful hum we hear, as the bee flies past us, does not, I think, sound at all like such a grumble. Do you think it does ? It may be the same thing every day, but it is what the bees have to do, and they do it very cheerfully ; and I am sure of this, that they are never so happy and in such good temper as when they are at work ; and never so cross, as you will find when you keep bees, as when they are obliged to stay at home by the weather being cold or wet. Then they are much more inclined to sting.

Their patience puts me in mind of a well-known patient donkey. At Carisbrooke Castle, in the Isle of Wight, where Charles the First was confined as a prisoner in 1647, there is a very deep well, three hundred feet deep, and, in order to draw the water, there is a contrivance of a great wooden wheel, which, when it is turned, draws up the bucket. This wheel is made so large and broad that a donkey stands inside, and turns it by stepping on, as if walking, although, in fact, the poor animal never really advances an inch, for, as it moves, the wheel of course moves from under its feet. What dull work does this seem, always stepping on, but always in the same place! But the donkey, like the bees, is patient. One donkey was known to do it for fifty years, and another for forty years.

CHAPTER V.

COMMUNITY OF BEES IN A HIVE.

THE next thing to notice, as we see the bees in hundreds going in and out of the same little door, is the fact of their living and working together, and helping one another. They form thus, what is called a community or colony.

In thus living together they are different from most insects and animals. Indeed, but few do the same. We may find in many cases vast numbers of insects living together in the same place, such as swarms of gnats in a damp cellar, or millions of flies filling the air, or coming in great numbers into a house. Or again, amongst birds, we may find thousands of starlings congregated together; or large colonies of rooks, many of them building their nests in the same tree, and then in winter time coming home to roost at night in countless numbers, so that the very air is darkened. But, although they thus congregate together, they do not form a community; they do not work together for a common purpose; they do not feed and take care of each other's young. Each insect looks out for the supply of its own needs. Each pair of birds,—sometimes the hen bird alone,—build their own nest and rear their own young, and have no regard to or interest in others.

But, when we come to look at bees and some other insects, we find a different state of things altogether. We find that everything, even their very lives, depend upon their living in a community or society, all obeying, by instinct, common rules, each one doing its own part in the common work.

> 'Alike ye labour, and alike repose;
> Free as the air, yet in strict order join'd,
> Unnumber'd bodies with a single mind.'
>
> EVANS.

We see the same, in great measure, in wasps, which live together during summer and autumn, all helping together in the work of the common home. We see the same in ants, which are insects in many respects as wonderful in their habits and instincts as bees.

Here, to illustrate what I have said about insects working together for a common purpose, I may relate a story told by Sir John Lubbock of some wonderful ants which actually make slaves of other ants, and, in order to obtain and bring them into captivity, go out on regular slave-making expeditions.

One day a whole body of these Amazon or slave-making ants was seen making its way, like an army of soldiers, all drawn up in battle array, and without straggling, across some distance of ground, and through a thick hedge, and straight on, until at last they reached the nest which they were intent on robbing. Then for a few minutes there was a fierce battle, but the Amazons soon got the best of it, and, forcing their way into the nest, were presently seen marching home, but each ant now carrying in triumph, as the spoils of victory, one of those little white things, often called ants' eggs (which however are really insects in a more advanced state), and which in their captors' nest soon would become live ants, and very useful slaves. And so you see instinct taught them to go out with a common purpose, to work together, and to assist one another.

I can tell you, however, a much more pleasant story given by the same author. I may say that in order to observe their habits, he kept a considerable number of nests of ants in his own house in little cases or boxes, made partly of glass, so that he could see all they did.

On one occasion, in one of his nests, there was a poor ant which, on account of being deformed, 'never appeared able to leave the nest.' However, one day, he says, ' I found her wandering about in an aimless sort of manner, and apparently not knowing her way at all. After a while she fell in with some specimens of the little yellow ant, who directly attacked her. I at once set myself to separate them, but, owing either to the wound she had received from her enemies, or to my rough, though well-meant handling, or to both, she was evidently much wounded, and lay helpless on the ground. After some time another ant, but from her own nest, came by. She examined the poor sufferer carefully; then picked her up gently, and carried her away into the nest. It would have been difficult for any one who witnessed this scene to have denied to this ant the possession of humane feelings.'

Again he says, 'At the present time I have two ants perfectly crippled, so that they are quite unable to move, but they have been tended and fed by their companions, the one for five, and the other for four months.'

See, then, how they not only live together, but are kind to one another, and help one another.

Beavers, again, amongst animals, are striking examples, in some respects, of the same thing. Their wonderful houses, built with rooms and passages, and made strong and secure with wood, stones, and mud, are made by them for the common purposes of the whole colony. In it they live and work together.

In the case of bees, this community or society is absolutely necessary. A single bee cannot live by itself. If you were to take a bee, or, we will say, half-a-dozen bees, and put them by themselves into the most comfortable little hive possible, they would very soon die. They would have no spirit to work. They would not even care to get food for themselves, although there might be plenty near at hand.

But how different is it when the whole colony is together! Then, by common instinct the bees seem as one united band of hearty, contented workers; working together for their common wants, helping one another whenever and however they can, each doing its own part, always happily at peace amongst themselves. What a good example do they give us!

THE SONG OF THE BEES.

Flying out, flying in,
Circling the hive with ceaseless din,
Now abroad, now at home,
Busy through wood and field we roam.
Here in the lily cup, there in the clover,
Gather we sweets the meadow over.
Food to our young we carefully take;
Pollen we bring, and wax we make;
A band of us shapes each tiny cell,
Another follows, completing it well.
Working all, working ever,
Suffering idlers among us never,
Never pausing to take our ease:
Oh, busy are we, the honey-bees!

Flying out, flying in,
Circling the hive with ceaseless din,
Now abroad, now at home,
Cheery we stay, and gaily we roam,
Never too hurried to greet a brother,
With feelers crossed we talk to each other;
Never too selfish to share our stores;
Some seek them abroad, some use them indoors;
Unitedly guard we our homes from harm,
Stationing scouts to give the alarm.

So, working all, and working with will,
Providing in summer for winter chill,
Whirring and buzzing, nor caring for ease,
Oh, cheery are we, the honey-bees!

Flying out, flying in,
Circling the hive with ceaseless din,
Whether abroad, or whether at home,
Loyal we stay, and loyal we roam.
In royal apartments our queen-bee is reigning:
We render our homage unmingled with feigning:
Lowly we bow as we pause by her side,
The choicest of food with her we divide.
Thus working all, and working with heart,
Each striving good to the whole to impart,
Busy and cheery, we think not of ease,
And loyal are we, the honey-bees!

Flying out, flying in,
Circling the hive with ceaseless din,
Whether abroad, or whether at home,
This lesson we teach wherever we roam:
Mortal, like us, go labour unwearily,
Work with thy kind, and work with them cheerily;
Duty fulfil, wheresoe'er thou may'st owe it;
Where honour is fitting, fail not to bestow it;
It matters not whether at home or abroad,
Be faithful to man and be loyal to God.
Thus work thou well and work thou ever;
The lessons we teach thee thou may'st not dissever:
Be busy, be cheery, be loyal, for these
Are the truths thou may'st learn from the honey-bees!

Child's Companion.

CHAPTER VI.

DIFFERENT KINDS OF BEES.—HUMBLE BEES AND THEIR USE.

HITHERTO we have only been speaking of bees in very general terms—the common hive-bees that we see working in our fields and gardens; but there are many other kinds as well, and, if we only use our eyes, we shall soon see some of them. To find specimens of them all, however, would be impossible, for there are more than two hundred and fifty different kinds in our country alone, and some of them are very scarce, and many of them are only found in particular places.

One kind of wild bees, however, you will certainly find without difficulty—at least, in summertime. I mean the large humble bees, which make such loud noise as they fly amongst the flowers, or when by chance they come into the house.

And then, besides these very large humble bees, you will soon find many others of different shapes, sizes, and colours—some of them very small. And as you look along some dry, warm bank, you will probably find the home of some one of these many kinds. You will see a very small hole, and some of the humble bees going in and out. And, if you follow this hole a short way, you will find the nest beautifully made; although not made to last through the winter, but only for the time necessary for the young bees to come to their full growth. Before the winter comes, and the banks become sodden with wet and snow, and the nests are thus destroyed, the young queens, or mother bees, leave their summer homes, and hide up for the cold months in some dry nook, or crevice of a tree, and only come out again, to begin work and to make their nests, when another year has come round, and the weather is fine and warm.

Now you have doubtless been taught that everything that God has made is for some wise purpose, and does some good, although in many cases we may not know what it is. And the more we observe all the habits of animals and insects, the more we shall see evidence of this great truth, 'Nothing without purpose.'

I mention this here because we see it remarkably the case with the humble bees. They have always been favourites—considered pretty harmless insects; but it is only quite lately that, by close observation, it has been discovered that they are of the greatest use, and do a most important work in our fields. I cannot now fully explain it all; but, to give you some idea of this discovery and its value, I may say that just because their tongue is a little longer than the tongue of other bees, they are so very useful amongst certain flowers (especially the red clover) in our fields, that a great deal of trouble has been taken, and a great deal of money spent, in order to send some of them all the way across the sea to the other side of the world—to Australia and New Zealand, where they were not found before. It is very difficult and expensive to send them so far; but they are so

Wild Bees and Flowers.

very useful to the farmers in those distant lands, that it is quite worth while. Hundreds have been taken, and let loose in the fields.

And then besides the humble bees there are, as I have said, many other kinds, some of which are termed 'solitary bees'—bees, that is, which live a solitary life; do not live in communities, but make nests by themselves and for themselves alone. Amongst these there is, for instance, the leaf-cutting solitary bee; which makes its little nest in the ground, or in clefts of walls or trees, with small pieces of leaf cut and fitted in with great care and trouble.

There is also the mason bee; so called because it builds its little house of small stones—or, rather, grains of sand—and plasters all, like a mason, with a kind of cement or mortar of its own manufacture. You may sometimes find one of these little nests, almost the size of a walnut, fastened on to an old wall; and so firmly made that a knife will hardly cut it. Or sometimes you may find them in very odd places indeed. I know of a case where the little bee chose for its nest the lock of a table drawer in a clergyman's study, and another the padlock of a door. These locks were found full of sand and dirt, and were at first supposed to have been injured in mischief; but upon being opened, were found to contain the nest of a mason bee with food for its young.

Another kind of bee I saw lately making its nest in an old nail-hole in the door of a shed. It was filling it quite full with food ready for its young.

CHAPTER VII.

VARIETIES OF THE HONEY-BEE.

ALTHOUGH we find the same animals in very different countries, yet, generally speaking, we find them varying in many respects—in appearance and habits—according to the country in which they live. Thus there are Arab horses and English horses, and, again, English dogs and French dogs. And in the same way, there are English bees, and Italian, Syrian, and Cyprian bees; also Indian bees and a race of stingless bees in Brazil, and very many more.

Again, just as some of the foreign animals are more valuable than the English varieties—as, for instance, the Syrian sheep, of which probably you have seen pictures with its long tail of valuable wool supported on a little carriage; so some kinds of foreign bees are better and more useful than the English, although we must add that some of them are bees of quick temper when carelessly treated, and sting very sharp indeed. With some, however, it is just the reverse, and this is especially the case with the Italian, or, as they are sometimes called, Ligurian bees; of which kind you will hear a great deal, as numbers of them are now kept in all parts of the country instead of the common English bees.

These Italian bees came at first from the north of Italy, and are exceedingly beautiful bees, marked with three bright golden bands or girdles; and are said to be the best-tempered and gentlest of all bees, so long as they do not mix too much with their English neighbours. But that which makes them most valuable is not their good looks, but their activity and industry. They are early risers, and will be at work before the other bees are out of their hives; and will continue to work in the fields and gardens later in the evening. They will also work longer into the cold weather of autumn, and at other times when most bees keep within doors. This is a very good character to give them—is it not?—early risers, hard workers, good-tempered. They are, I think, quite the sort of friends we should try to make.

Italian Bee.

Then, again, amongst the other varieties of which I spoke, there are some, of which probably many more than at present will be kept in England before long. The Syrian, for instance, is a very valuable race of bees. They are smaller than the Italian, but are marked in very much the same fashion. Unfortunately, however, they are very bad-tempered. This also is the character of the bees from the island of Cyprus; which, however, notwithstanding their angry disposition, some say are the best of all bees.

A well-known bee-keeper went to Cyprus in 1882, taking the long voyage for the purpose of bringing home to England a great many of these bees. He tells us how, after much trouble, he bought forty hives in one place, and carried them a long way over rough mountain roads, on the back of mules, each mule carrying two colonies in the earthen hive of the country, slung, one on each side of the mule. On one occasion, however, the bees quite lost their temper. Perhaps he shook them, or disturbed their homes in too rough a manner; and then, to teach him to be more gentle and careful, they punished him with a hundred stings.

If we go to India, we find many other kinds. The largest honey-bee yet discovered is a native of Hindostan, Ceylon, and the Malay Peninsula. It collects immense quantities of honey, which it stores in huge combs suspended from the topmost boughs of the tallest palm-trees, and also from rocks, in places often far out of reach. Some people have tried to keep them, but have not as yet succeeded, for the race is a very wild and savage one, as much so, in their way, as the terrible tigers of the country.

CHAPTER VIII.

AMERICAN BEES—THE BEE-LINE.

AMERICA, as well as other countries, has its own bees. You will hear something later on of bee-keeping in America, and the vast scale in which it is carried on there; but at present we are thinking only of the wild bees, or native races. There are a great number of these inhabiting the extensive forests of the country. And it is said that where monkeys abound, the wonderful instinct of the bee teaches it that the only safe place in which to build its nest, in order to be out of the way of these active thieves, is on the topmost and most slender boughs of the trees, where even a monkey cannot climb.

In some parts of the country, where the nests are in hollow trees, or any other accessible place, a bee hunt often affords great amusement as well as profit. The hunter goes out near the woods, and, after catching a bee, gives it as much honey as it can eat and carry; and then, getting himself into a good position, so that when the bee flies he can see its little form against the light sky, he lets it go. The bee, after making a circle or two, goes straight home, the man watching it as far as he can, and taking particular notice of the direction in which it goes. It soon comes back again for some more honey, and the hunter knows it to be the same bee, for he has marked it with a little red paint.

Again he feeds the bee as before, and then, going in the direction he saw it take the first time, he lets it go again, and marks its flight. And so, by degrees, he gets nearer and nearer to the nest.

Then he takes his bee, and goes to the right or left of the line, and lets it go again. Straight it flies, making thus, of course, a new 'bee-line,' as it is called, at a certain angle to the first line. Observing this carefully, the hunter knows that where these two so-called lines meet one another is the exact spot where he will find the nest. So it proves, and he takes the honey. It requires, of course, much care and ingenuity, but in this way affords good sport as well as profit.

You may try and find a wasp-nest some day much in the same way, for wasps, as well as bees, fly in a straight line when returning home. There is no loitering idly, remember, on the way, as very often we see in the case of boys and girls when sent on an errand—stopping and playing by the roadside, and forgetting for a time what they have been sent to do. No; the bees go straight, and go as fast as they can. They have their work to do, and they do it.

How they are able to make this straight 'bee-line' home, even when they have never been the way before, is a great mystery. It is, indeed, by what we call their instinct, although we little know, perhaps, what instinct is. We only know that it seems in some way a marvellous power given them by the Creator, which, in many respects, almost supplies the place of the powers of reason given to man, and often enables them to do what man with all his reason never could.

It is the same instinct which is found even yet more wonderfully in some animals, and especially in dogs, who will find their way home for one or even two hundred miles across a strange country, where they have never been before.

A cat will sometimes do the same. The following story was given me on good authority:—A cat was taken by a lady from London to Lowestoft, on the Suffolk coast, by railway, a distance of a hundred and eighteen miles. There it escaped, and in a fortnight's time appeared at its old home in London, having found its way by the teaching or leading of its instinct. This is, indeed, far more wonderful than what bees can do, but it is example of the same kind of instinct.

CHAPTER IX.

BEES IN THE OLDEN TIME.

BEFORE we think more especially of English bees and bee-keeping, it will be interesting to look into some of the records of the long-ago past, and to see what was known of bees in the earlier ages of the world, and how far they were valued.

With this object in view, we look first at the Bible, and there again and again, in almost all parts, we find some mention or allusion to bees, or honey, or honeycomb. And we are led to think that, as in these days, the Holy Land had a very valuable race of bees, which greatly abounded, and gave honey held in high estimation and largely used as food.

In the very early days of the Patriarchs we know that the honey of the country was esteemed of sufficient value to form part of the 'present' which Jacob sent down into Egypt by his sons to appease the ruler of the land, his own son Joseph, that so he might send away his other son and Benjamin. The 'present' was a 'a little balm and a little honey, spices and myrrh, nuts and almonds.' Again, in Ezek. xxvii. 17, we read of honey as a distinct article of 'trade,' mention being made of Judah and the land of Israel trading in honey with Tyre.

Again, we read in the Bible of bees, just as in these days, building their nests in very various places —rocks, trees, and so forth. The Psalmist speaks (Psa. lxxxi. 17) of 'honey out of the stony rock.' And it was in 'the wood,' when 'honey dropped' from some nest built on a tree, that Jonathan took a little to satisfy the cravings of hunger, and without knowing it, disobeyed his father's command. And then we read of a colony of bees which actually made its nest in the carcase of the lion which Samson had killed some time before.

Whether the bees were, in any way, kept in hives, or the honey simply taken from wild bees, we can hardly say; but, whatever the case in the Holy Land, bees were certainly thus kept (and had been so for long) in other countries in the time of our Lord. John the Baptist in the wilderness ate 'wild honey,' implying, perhaps, that some honey was to be had from bees not in a 'wild' state. At all events, in Greece and Italy bees had both been 'kept' and observed long before this time.

Among the many who wrote of bees and honey in those olden days, Virgil, the great poet of Italy, who lived and died a few years before Christ, stands first of all. He devoted the whole of one of his books to the subject; and although he made a great number of strange mistakes, and took many of his ideas from yet more ancient authors, and probably was not himself a bee-keeper, he must nevertheless have taken considerable interest in the subject

As example of his errors, or of the common ideas of those days with regard to bees, he supposes that when bees are lost, a fresh colony can be obtained from the carcass of a young ox; and he gives many and exact directions how to proceed in such a case.

He also speaks of bees carrying little stones to serve as ballast to steady them in stormy weather:—

> 'And as when empty barks on billows float
> With sandy ballast, sailors trim the boat;
> So bees bear gravel stones, whose poising weight
> Steers thro' the whistling winds their steady flight.'

On the other hand, he gives some directions as to bee-keeping which are excellent, especially as to the situation for an apiary—with sun and yet shade, sheltered from winds, and with some water near at hand.

Less than a hundred years afterwards Columella lived and wrote on the same subject, and others also, but not with greatly increased knowledge.

We then hear but little of the subject until about a hundred and fifty years ago, when the whole study of natural history revived.

We must, however, pass over this period, for I want to point you especially to one great observer and writer about bees who lived about a hundred years ago—Huber; and I want to do this because, when thus an observer and writer, he was totally blind. Think of a man who was quite blind taking an interest in bees, and knowing a great deal about their habits, and finding out very much that had never been known before! Does it not seem very strange and wonderful?

Huber was born at Geneva, in 1750. At an early age, when little more than a boy, his eyesight greatly failed, and he was told the sad truth that in a little while he would for ever lose the precious gift. Like a man of true courage, he did not, however, lose heart, but determined with himself that, although in darkness, he would try to live and act, as far as possible, as if he could see. It was a noble resolve, and had its reward.

In his early boyhood he was fond of natural history; and having, after blindness came on, been led by the writings and conversation of a friend to take an interest in bees, he set himself with all the zeal and energy of his nature to study them for himself, and, from that time forward, devoted himself, almost entirely, to examine into some of the most difficult questions connected with their habits and natural history.

The story of his observations, discoveries, and various ingenious experiments, is most interesting, and you will do well to obtain his biography, and read it. Much, however, that he did would have been impossible had it not been for an excellent and devoted wife, who for forty years never ceased her loving and attentive care, but in every way sought to lighten his affliction, and to help him in his work, reading to him, writing for him, and, as far as possible, giving him the use of her own eyes. He used to say of her 'that as long as she lived I was not sensible of the misfortune of being blind.'

Huber had also a most useful and intelligent servant whom he trained to be a very close and exact observer, and whose eyes he thus used instead of his own.

Their patience in observation and experiment was most remarkable. On one occasion they looked at and examined every single bee in a hive to find out something they wanted to know. At another time for days, and perhaps months, they would watch, observe, and make experiments to discover, if possible, the truth respecting some one little thing which they did not understand. At another time they would invent some clever contrivance by which they could see exactly what the bees were doing inside the hive.

But I cannot now tell of all these things. I now chiefly point you to Huber, not only as an observer of bees, but that you may see in him an example of courage under difficulties, and how patience perseverance and ingenuity can accomplish great things; and how it is possible, even with such an affliction as total blindness, not only to be resigned, content, and happy, but also to live a life of usefulness.

CHAPTER X.

THE INHABITANTS OF THE HIVE.—INTRODUCTORY.

SINCE Huber's time great advances have been made in the knowledge of bees, as of everything else. It has been, as we know, an age of discoveries. Steam, for instance, and its marvellous powers, applied to railroads, machinery, and ships, has brought about a change which to our forefathers would have seemed an impossible dream. The electric telegraph brings people far away into instant communication with one another. And every day fresh discoveries are made by those who carefully study and observe.

And as with great things, such as steam and electricity, so it has been with the little subject of bees. Many things are known now which were not known a few years ago, and fresh things are being found out even now continually; and everything that is so discovered makes their history, habits, instincts, and uses, appear more and more wonderful, giving us more and more insight into the marvels of creation, and making us feel all the more the truth of what the Psalmist says, 'How manifold are Thy works! in wisdom hast Thou made them all.'

In order to understand some of these discoveries you will have to give much attention; for the lessons about them will not be altogether easy; but at present we will only think of simple things.

And first of all, we will go to a hive, standing in some garden near at hand, and ask the bees to tell us a little of their history, taking care to go to them quietly and to treat them with gentleness. And our first inquiry must be this: 'Who is at home?' and, in the next place, 'What have you got inside your hive?'

To these questions what answer shall we get? Well, it will a good deal depend upon the time of year, both as regards the number of inhabitants and the description of bee, as well the contents of the combs; for, in the first place, whether the hive is one of English bees, or one of Italian, Cyprian, or any other race, we shall always in summer time—if it is a healthy hive—find three kinds of bees; but at other times only two.

Dealing, first, with the bees themselves, the two kinds always present are, (1) the queen; and (2) all the common workers; and then we have, thirdly, the drones; but these last are only found in summer, or rather from about May to August.

Here is an illustration of each kind :—

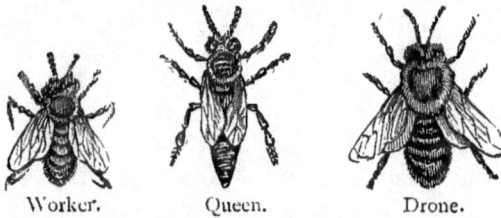

Worker.   Queen.   Drone.

At a future time we will think of, and look closely at, their wings, legs, and stings, and some of their other parts, but at present let us only take notice of their chief features.

I. The Worker Bee.—This is the common bee, which most thoroughly deserves its name, and which you know so well in appearance, although perhaps you have never stopped to inquire how many legs or how many wings it has.

You will notice that it is the smallest of the three kinds. These 'workers' are all female bees, and are sometimes called neuters,—a name given them because, although females, they never, or only very rarely, lay any eggs. And it is much better for the hive that they should not do so. Indeed if one of them does lay eggs it generally quite spoils the whole hive.

There are a vast number of these workers in a hive,—as many as from 15,000 to 40,000, or even perhaps 50,000; as many bees, that is, as there are people in a very large town.

II. The next illustration is that of the Queen, or Mother bee, who reigns by herself, the only one of her kind in the hive, chief of all, and most important of all.

> 'First of the throng and foremost of the whole,
> One stands confest the Sovereign and the soul.'
>
> VIRGIL.

Now observe her appearance carefully, for, when you keep bees, you will have to be perfectly familiar with the appearance of a queen, and to be so quick with your eyes as to be able to find her out amidst the thousands of others. You will learn to do this without much difficulty after a time. I can fancy I now hear you saying with delight, 'There she is! there she is!' as you point to her walking about amongst the crowds of others on all sides, and surrounded by her attendants, every one with its head towards her majesty, stroking her, feeding her, and following her wherever she goes.

And now, observing her closely, we notice, first of all, that she is a much longer bee than the others, and that her body is of an elegant shape, and that, comparatively to her body, her wings are much shorter. We notice also that her colour is somewhat different to the others, rather darker and brighter.

Observing her movements, we notice that she walks about more slowly and sedately than the other bees, as well becomes her position of queen, or rather, we might say, the mother bee of the hive.

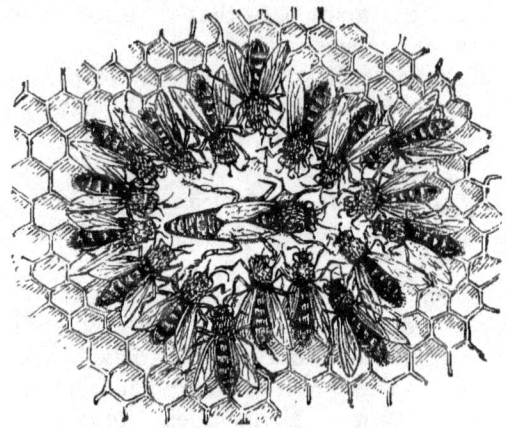

The Queen and her attendants.

'Twelve chosen guards, with slow and solemn gait,
Bend at her nod, and round her person wait.'
                                                    EVANS.

She is always called the queen, but this term of mother bee' is perhaps the most correct; for this is what she really is, the mother of all the bees in the hive, the honoured and respected head of the whole family, both workers and drones. And it is most interesting to see the care the workers take of her, and how they treat her, as a mother ought always to be treated by her children; how they wait upon her, and provide for all her wants, and mourn for her if she dies; and indeed soon themselves pine away and die, unless they can get another to take her place.

In olden days it was not by any means universally

known that the queen is thus the one mother bee of the hive. Virgil, amongst his many strange mistakes, speaks of her, not as a female at all, but as a king; and, when he describes the battle of the bees, speaks of two kings leading forth their hosts to war, and themselves joining in the fight.

> 'With mighty souls in narrow bodies prest,
> They challenge and encounter breast to breast.
> So fixed on fame, unknowing how to fly,
> And ultimately bent to win or die;
> That long the dreadful combat they maintain
> 'Till one prevails, for one alone can reign.'

And, more or less, this mistake as to the queen's sex continued to the time of Shakspeare, who, about three hundred years ago, wrote of bees,—

> 'They have a king, and officers of state.'

This error is the more strange, because long before Virgil's time, the truth was known to some. Aristotle, who lived even three hundred years before Virgil, writing of bees, tells us: 'Some say that the rulers produce the young of the bees.' And again: 'There are two kinds of rulers; the best of them is red, the other black; their size is double that of the working bees. By some they are called the "mother bees," as if they were the parents of the rest.'

And in the time of Shakspeare, Dr. Butler, one of the first English writers on the subject, had some knowledge of the truth, although his idea was that the queen only laid eggs producing queens, and that the workers—known to him as females—laid all the other eggs. The full truth, indeed, was hardly known until the time of Huber.

III. But besides the one queen and the many workers, there are the drones, which—as was mentioned—are only to be found in the hive in the summer months; and of these there are, perhaps, 500, or sometimes as many as 2000 or 3000.

Look at the illustration of the drone—page 36—and you will notice that it is altogether a larger bee than the workers, and of different shape—very stout, broad, and bulky, and that its wings are large. These drones are the male bees of the hives and a very idle set they are, not at all deserving the name of 'busy bees.' To hear their loud hum, and the noise they make as they fly out on some sunny day, one might think they were doing a great deal; but if we go to the flowers we shall not find them there. In fact, they never do any real work, and are such helpless bees that they do not even get food for themselves, but live upon what the workers bring home. Shakspeare rightly calls them

'The lazy, yawning drone.'

They are, I think, very like many people who make a great fuss and loud boasting, and try to attract attention, and yet do not really do half so much work as those who make no pretence but go about their work, whatever it is, quietly and steadily, without noise or boasting.

'Buzzing loud,
Before the hive, in threat'ning circles, crowd
The unwieldy drones. Their short proboscis sips
No luscious nectar from the wild thyme's lips;
On others' toils, in pamper'd leisure thrive
The lazy fathers of th' industrious hive.'
EVANS.

CHAPTER XI.

THE HOME OF THE HONEY BEE.—INTRODUCTORY.

IN the last chapter we considered a few simple things, in a general way, about the inhabitants of the hive,—the queen, the workers, and the drones. Leaving for the present the consideration of what is more difficult to understand about them, we will now, in the same way, try and get a general idea of the wonders of the hive itself—the home of the bees.

And looking into a hive, the first thing we notice is, of course, a number of combs, of which you know well the general appearance. If we are examining a common straw hive, we shall see that these combs are of different sizes and shapes, all made to hang from the top of the hive, and so carefully arranged side by side, that just sufficient space is always left, between any two of them, to allow the bees, when crawling about them, to pass one another easily,—

'Galleries of art, and schools of industry!'

And now—as what we find in the hive will depend in some measure on the time of year—let us first of all suppose that it is summer time. Let us say that it is the month of June. And then, when we examine the combs, we find in the centre of them all—in the best and warmest part—the portion which is called the 'brood-nest,' or, as we may term it, the nursery of the hive, where there are in the cells great numbers of young bees in all the different stages of insect infancy, all being cared for with the greatest attention until they are fit to provide for themselves. We shall find this state of things more or less all through the year, except in winter; but at no time more than in June.

This brood-nest will generally occupy the greater part—at least all the middle part—of several of these central combs. In the outer portions of these combs, which are not suitable, being too chilly, for the young bees, there will generally be honey safely stored away.

In the brood-nest itself we shall find some cells closed and some open.* In the open cells we shall see either a very small white speck, which is an egg, fastened to the bottom of the cell, or else what appears like a little white maggot; some of these latter will be small, and some of good size, nearly filling the cell.

Of the closed-up cells some will appear with a flat dark covering; and out of these will soon come the perfect, full-grown young worker bees. Others will appear—but we shall not find them in every comb of the brood-nest—with a much higher and rounder top. Out of these will come, in due course of time, some of the big, idle, noisy drones.

And then, finally, on this comb from the brood-nest that we are examining, we may or may not find (when present, we shall find it generally on the edge of the comb) a large, dark-coloured cell, in appearance like an acorn, hanging by itself; and if so, then inside it there is a young queen. It is a queen-cell.

You must remember, however, that this state of things in the brood-nest is the condition of summer-time. If our visit to the hive is in winter, we shall

* See illustration on page 137.

not find eggs or young bees, but we shall see in the brood-nest and the adjacent parts, all the bees, as far as possible, huddled together to keep themselves warm.

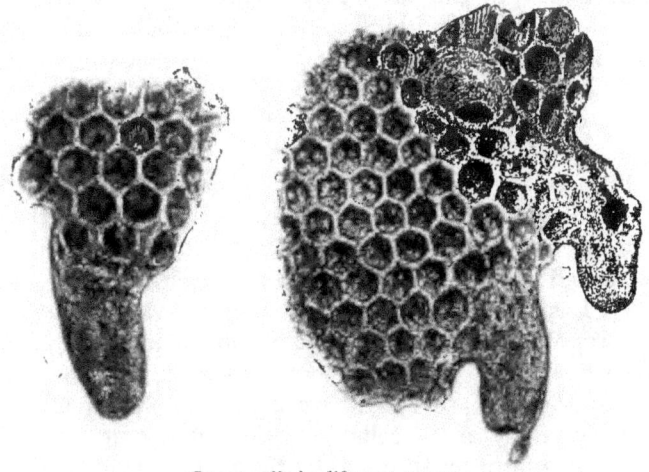

Queen-cells in different stages.

All the combs, not required for the brood-nest, may be considered the great store-room of the hive in which the bees keep all the food they are likely to need at a future time. A great portion of the brood-nest itself they also use for the same purpose, after the breeding season is over, and the cells are no longer needed for young bees.

And now, what are these stores? First of all, of course, there is the honey,—not much in the early spring, but more and more as the year gets on, until at last almost every cell is full, and ample provision has been made for the winter supply of the hive.

It is not, however, honey alone that the bees store away. In many cells we shall find the substance called 'pollen,' which is the food of the infant bees, and without which they cannot thrive. We find it, especially, in the early part of the year when many young bees are daily coming into the world, but, more or less, at all times.

It is sometimes called 'bee-bread,' and appears in the cells as a sticky, and rather hard, substance, and is made,—as you will hear in a future chapter,—from the little yellow pellets which you must often have noticed sticking to the hind-legs of bees, and which, when they bring home, they mix with a little honey, and, if not wanted at once, put away into the cells for future use.

Lastly, in the hive we shall find what is called 'propolis,'—a very thick sticky substance, which, after a while, gets hard like cement. We shall not, however, find much of this; and we shall not find it stored away in any of the cells, but only put into cracks and crevices in order to make all tight and secure, and to shut out cold and draughts.

Thus in the hive we find a palace for the queen, a nursery for the young ones, a store-room for food, and a comfortable home for all.

As we take such a general view of the interior and its contents, one of the many things which will strike us very much, will be the wonderful way in which the combs are all made to fit into the space which the bees have at their disposal, and how they are contrived so that no room is lost. If one comb is a little twisted, the next one to it is made with just the same twist; and if there is a little vacant corner anywhere, a little bit of comb is made exactly to fit it, and very often in the most curious and ingenious way possible.

There is no waste of room or material, or carelessness as to little things. If we could hear them talk we should never hear them saying, 'Oh! never mind that little bit of wax, or that little corner of the hive. It will not much matter if we do waste it. It won't make much difference. It's only a trifle.' And if they meet with difficulties, and get into awkward places, as they often do in badly made hives, or in trees or buildings, they will always, in a most wonderful way, make the best of the situation, and adapt themselves to circumstances. We cannot do better, I think, than try and follow their example.

Another thing that will certainly strike us will be the tidiness of everything,—the whole house kept in good order. We shall see an excellent example of the old saying, 'A place for everything, and everything in its place.' We shall not find dirt, dust, and refuse left about, if only the weather is such that they can get rid of it. If there is a piece of dirt, or a dead bee, we shall see them pulling at it with all their strength; and if it is too much for one to manage, we shall see two or more joining in the work, until they get it out of the hive and throw it on the ground.

It is quite the tidy house one likes to see,—everything clean, even if old; everything in its place, and everything well ordered, and done at the right time. It is not the home one often sees,—without order or arrangement, dirty and uncomfortable, and everything in confusion from morning to night.

The Cottager and his Bees.

CHAPTER XII.

THE HISTORY OF THE HIVE.

The colony of bees, described in the previous chapter, is one that is in a prosperous condition. There are plenty of bees, and the hive is full of comb; and there is abundance of brood and plenty of honey, and pollen stores, and all things in order. In the next place let us see how all this has come about. What has been the history of the hive? This, to some extent, depends on its age; but as, after a colony is well established in the hive, its history is much the same every year, we will think of it especially in its early days—the first year of its history.

And very possibly, although we see it now so full and prosperous, it is not more than a few months old. It is now, let us say, the month of August, and very possibly it was only last May that the bees first took possession of the hive.

To trace its history let us go back in thought to that merry month of May, and I will suppose that you and I are together, amongst the hives, on some bright morning of that beautiful month, when all nature seems to be putting forth its freshest vigour; and we stand and admire the lovely sight of the orchard trees,—the apple, pear, cherry, and others,—full of blossom; while the bees from all the hives fill the air with their pleasant hum.

And now I call your attention to what appears as an unusual state of things at the entrance of one of the hives, and we notice that the bees are evidently not working as usual. They seem restless and excited, flying round and round, and not going far from home. The entrance is especially crowded. Possibly there are numbers of bees hanging together there in a great cluster. The great drones also partake in the general excitement in their own noisy way, rushing in and out, and circling round, as if determined to be seen and heard.

The fact is, the bees are preparing to swarm. Let us observe them closely. And we have not watched them long before we notice hundreds of bees, perhaps very suddenly, pouring out of the hive, and hundreds more pressing after them as fast as they can get out of the entrance, tumbling over one another in their haste, and then flying round and round; and more pressing out, until the whole air seems filled with bees in the most excited state.

> 'Upward they rise, a dark continuous cloud
> Of congregated myriads numberless,
> The rushing of whose wings is as the sound
> Of a broad river headlong in its course.'
>   SOUTHEY.

But before they do all this I say to you, 'Let us venture near the hive; and you need not be afraid to do so, for bees, when they are swarming, are generally in the best of tempers. Let us watch the entrance, and, perhaps, we shall see the queen herself come out. Yes, look! there she is! Do you not see her? She joins the throng, looking quite the queen, the acknowledged ruler amongst her many faithful subjects. And now, as we watch the crowd in the air, we notice that many of them seem to be gathering round, and settling on one of the boughs of a neighbouring tree. It evidently is so; and then others follow to the same place, and more and more collect, clinging to one another, until at last they hang down from the bough in a bunch larger than a man's head, and even bend down the bough with their weight.

> 'Round the fine twig, like cluster'd grapes, they close
> In thickening wreaths, and court a short repose,
> While the keen scouts with curious eye explore
> The rifted roof, or widely gaping floor
> Of some time-shatter'd pile or hollow'd oak,
> Proud in decay, or cavern of the rock.'
>   EVANS.

And now, as you look at this swarm, you would doubtless like to know how many bees there are in it. How many do you suppose? Well, if we were to weigh the whole lot we could almost tell the number; for 30,000 bees, generally speaking, weigh rather more than 6 lbs. (6 lbs. 5 oz.), and it is possible there may be this number in the swarm we are looking at; but, if so, it is a very large one. An average swarm contains about 15,000 or 20,000 workers, besides several hundred drones, and, of course, the one queen. I am speaking, however, now of what is called a first swarm, for it is rather different in a second, or what is usually termed a 'cast.'

But we must not stand watching the swarm too long, for, unless we take measures to secure the bees, they will fly again, and take possession of the 'rifted roof' or the 'hollow'd oak,' and we shall lose them altogether.

In order to secure them, I get an empty hive; and I take this, and, placing it under the hanging cluster, shake the bough, when all the bees drop into the hive, which I then immediately turn over, and place upon a board on the ground, leaving a crack for the bees to come out and go in.

A vast number of bees at once rush out, but very soon, if the queen is in the hive, and the bees like its appearance, they settle down, and take possession of it as their new home, instead of the place they had found and intended to occupy. Thus the swarm is secured, and we carry it gently to its stand. Sometimes the swarm settles in a very awkward place,— very high up in a tree, where it can only be reached by means of a ladder; or sometimes the bees will settle round about the body of the tree itself, from which they can only be swept with a light brush, or gently persuaded to move by a little smoke, or the smell of carbolic acid. A little ingenuity, however, will generally very soon get over such difficulties.

Our hive being now in position, the bees at once begin, with the greatest energy, to make some comb in their empty, unfurnished house. They do not lose a minute, and they are able thus at once to begin comb-building, because they have been very provident, and have brought with them from the old hive as much honey and material as they could possibly carry.

Thus even in an hour's time they will have made a

Hiving a Swarm.

little bit of comb, and, in a very short time, will have made a sufficient number of cells for the queen to lay a few eggs. And then, if the weather is fine, so fast is the building proceeded with, and such numbers of eggs are laid by the queen, that in about three weeks' time many young bees are hatching out, and, soon after, hundreds more, and then day by day get greater numbers.

And all this goes on more and more rapidly as the young bees themselves join in the work of the hive, so that now, in August,—the time I named,—every corner of our hive is as full as possible of both bees and honey, and everything is prepared, and the whole colony is in a prosperous condition for the winter months. And, if we only keep it dry, it will be well able to stand the frost and snow,—the bees all huddled together, and keeping themselves warm, however severe the cold, until at last the spring-time comes again, and out-of-door work once more commences.

> ' Rous'd by the gleamy warmth from long repose,
> Th' awaken'd hive with cheerful murmur glows:
> To hail returning spring the myriads run,
> Poise the light wing, and sparkle in the sun.
> Yet half afraid to trust th' uncertain sky,
> At first in short and eddying rings they fly,
> Till, bolder grown, through fields of air they roam,
> And bear, with fearless hum, their burdens home.'
>   EVANS.

CHAPTER XIII.

A TALE OF DESTRUCTION.

HAVING described in the previous chapter the probable early history of the hive, which we saw so prosperous in August, I have now to relate a sad story of death—the end of many such hives. It is a tale of cruelty and improvidence.

You know the fable of the foolish man who possessed the wonderful goose which day by day laid golden eggs, and which would have enriched the man if only he had been content to wait for all the many eggs the bird would have given him. Impatient, however, to be rich, he killed the bird so that he might get at once all her golden eggs, but found, of course, that in doing so he lost everything, his bird and its eggs, and was left himself a poor man after all.

Well, in very much the same cruel, foolish way, it used to be the common practice everywhere in this country to kill the bees in order to get their honey, instead of preserving them to work again at a future time. And, I am sorry to say, this bad old custom still prevails in many places.

On some August evening, when the hives are full of bees and stores, and all are at home, ready for work again on the morrow, the bee-keeper (although bee-murderer would be a better name) comes in the dark to do his deed of cruelty, and digs a small round hole, at the bottom of which he places burning sulphur.

Then, taking the hives one by one from their

stands, he places them over the burning pit; where the horrible sulphur-fumes, rising up into the hive, soon destroy all life within; but not before you may hear a loud humming noise, the dying cries, as it were, of the thousands of bees as they fall from the combs into the pit below,—cries that seem loudly to reproach the cruel owner for his ingratitude to his faithful servants, rewarding them with death after they have worked hard, and done all they could for him, and were ready to do a great deal more.

> 'Ah! see where robb'd and murder'd in that pit
> Lies the still heaving hive! at evening snatch'd,
> Beneath the cloud of guilt-concealing night,
> And fixed o'er sulphur; while, not dreaming ill,
> The happy people, in their waxen cells,
> Sat tending public cares, and planning schemes
> Of temperance for winter poor; rejoiced
> To mark, full flowing round, their copious stores.
> Sudden the dark oppressive steam ascends;
> And, used to milder scents, the tender race,
> By thousands, tumble from their honey'd domes,
> Convuls'd and agonising in the dust.
> And was it then for this you roam'd the spring,
> Intent from flower to flower? for this you toil'd
> Ceaseless the burning summer heats away?
> For this in autumn searched the blooming waste,
> Nor lost one sunny gleam? for this sad fate?
> O man! tyrannic lord! how long, how long
> Shall prostrate nature groan beneath your rage,
> Awaiting renovation? When obliged,
> Must you destroy? Of their ambrosial food
> Can you not borrow; and, in just return,
> Afford them shelter from the wintry winds?
> Or, as the sharp year pinches, with their own
> Again regale them on some smiling day?'
>
> THOMSON.

But it is not only a cruel system, and a foolish one, because like killing the goose that laid the golden eggs; but it is also quite unnecessary, for, by proper management, all the honey can be obtained without killing a bee.

It is also a very profitless system as regards the honey itself, making it of a very inferior quality. And we can easily understand this when we remember that the hive not only contains much old comb, blackened by age and the impurities left by young bees, but also that in many of the cells there will be a quantity of pollen, or bee-bread, and even young brood in various stages,—in appearance like so many little maggots. All this comb is cut out of the hive, and broken up, and then smashed, and pressed, and mixed with the honey, before this latter is strained off. The honey thus obtained, and which many people eat, must indeed be very impure and inferior, and not for a moment to be compared with that which the bees will give us, clear, and bright, and clean, just as they themselves store it, if only we treat them properly.

How to obtain this pure honey, stored in perfectly fresh, clean combs, and in abundant quantity, and with considerable profit, you will learn at a future time.

Well, then, let this be your first great lesson in practical bee-keeping,—a lesson of what you are not to do: 'Never kill your bees.' Always look upon the sulphur-pit system as a most cruel and wasteful one,— the system by which you get the smallest and worst return possible for your money, time, and labour.

CHAPTER XIV.

INTELLIGENCE AND OBSERVATION NECESSARY.

I MIGHT now tell you of the new and better way of bee-keeping, and show you how even boys and girls may so keep bees as to find, in the occupation, both interest and amusement, and also earn something to put into a savings bank,—as bees put honey into their cells,—ready for a future time. But before I tell you of bee-keeping I want you to understand bees, and so must tell you something more than I have done yet, of their natural history, their habits and instincts; also of the structure of the several parts of their bodies, as, for instance, the head, the mouth, the wings, the legs, the sting; and how each is wonderfully made and fitted for the work it has to do. I must also tell you how bees differ from other insects, having certain habits, and certain parts of their structure, peculiar to themselves.

You must take trouble to understand something of all this, for the more you understand it, the more, I know, you will take interest in the subject, and love your bees. You will be able also to keep them better and with more profit, because you will manage them intelligently; you will know why you are to do this, and why you are to do that, and why you are not to do something else. You will see a reason for everything, and have a reason for all you do.

It is the same with bee-keeping as with every other occupation of life, you must understand first principles in order to do it thoroughly well; you must work with your head as well as with your hands. The gardener to be a good gardener, must be able not only to dig and root out weeds, but must understand something of the habits and growth of flowers, fruits, and vegetables; the farmer must know about the different qualities of land, and how to cultivate the soil according to its nature. Or, if a man is an engineer, he must learn by close study the nature of the materials, and the power of the forces with which he has to deal, such as the strength of iron and steel, and the true reason of why this or that is to be done.

This was how Huber made his great discoveries. He took the greatest trouble to understand even the most trifling things; nothing was overlooked. And it is the same with the study of every department of natural history. Observe everything. This is the foundation of success. A well-known naturalist has said, ' It is impossible to say at the moment of what use the most trifling facts may be. It is impossible to determine the exact importance of any circumstance in the history of an animal until we know its whole history.' And this is most true of bee-keeping. We shall succeed all the better by taking trouble to understand the bee, and by close observation of little things in its natural history.

In order to impress this truth, and to illustrate how great results may come from the exercise of such a habit, a few examples may be given from the lives of distinguished men.

It is said[*] that ' when Franklin made his discovery

[*] Smiles.

of the identity of lightning and electricity it was sneered at, and people asked, 'Of what use is it?' to which his apt reply was 'What is the use of a child? It may become a man!' Again, many before Galileo had seen a suspended weight swing before their eyes with a measured beat, but he was the first to detect the value of the fact. One of the vergers in the cathedral of Pisa, after replenishing with oil a lamp which hung from the roof, left it swinging to and fro, and Galileo, then a youth of only eighteen, noting it attentively, conceived the idea of applying it to the measurement of time. Fifty years of study and labour, however, elapsed before he completed the invention of his pendulum, an invention the importance of which can hardly be overvalued.

'Again, while Captain (afterwards Sir Samuel) Brown was occupied in studying the construction of bridges, with the view of contriving one of a cheap description to be thrown across the Tweed, near which he lived, he was walking in his garden one dewy autumn morning, when he saw a tiny spider's-net suspended across his path. The idea immediately occurred to him that a bridge of iron ropes or chains might be constructed in like manner, and the result was the invention of his suspension bridge.'

I can give you also another story, teaching us not to be satisfied until we know, if possible, the object and the use of even the smallest thing in any department of natural history, which we may be studying. It is the story of another discovery, as wonderful as any ever made, and which has brought the greatest blessings to the world, and has been the means by which, through the knowledge it gives of the secrets of life, innumerable precious lives have been preserved. I mean the discovery by the great Hervey of the circulation of the blood, or the way in which the blood is constantly, and every moment, flowing onward from the heart, through the arteries, to every portion of the body, returning by the veins to be first purified by the lungs, and then returned to the heart, ready to start afresh on its course, never ceasing till the moment of death.

This discovery is said by Hervey himself to have sprung from his seeking to know the use of some little valves which are found in the veins. They are, apparently, so insignificant that no one, before, thought much about them. Hervey knew they must have some use, and so set himself, with much study and by endless experiments, to find out the why and the wherefore. At last, after eight years, he was led gradually, step by step, to make his great discovery, which, although treated with scorn and all kinds of opposition at first, for ever marks him out as one of the greatest benefactors of the world.

CHAPTER XV.

THE NATURAL HISTORY OF THE BEE.

THE first thing, in the natural history of the bee, which you must, in some measure, understand, is the bee's position in the great insect world, or something of what is called the classification of insects.

As everything in a school would be in confusion without arrangement of the children into divisions, classes, and standards; so, to prevent confusion in the study of natural history; all animals are, as far as possible, arranged according to certain rules, each animal in its own proper place.

For this purpose animals are, in the first place, grouped together into certain great 'Classes,' such as the Class of Mammalia (animals that suckle their young), or the Class of Birds, or the Class of Reptiles.

These great classes are, in the next place, subdivided into large groups called 'Orders,' according to certain points of resemblance between the animals contained in any particular Order. For instance, amongst Mammalia there is the Order of flesh-eating animals, and the Order of animals like the ox, and the Order of monkeys, and so on.

Then, in the next place, these 'Orders' are divided into large 'Families,' according to yet further points of resemblance, such as the Family of all animals like the Cat, or the Family of all animals like the Dog.

In the same way these families are again divided into smaller groups called 'Genera,' and the genera into 'Species.' And of species there are Varieties.

And what has been done with animals has also been done with insects.

They are, in the first place, put in a Class; and are called Insects, because the whole body is, to a certain extent, divided, and consists of three segments, some being larger or more distinct than others.

You will see these parts very plainly in a bee.

There is first the Head. Then, very distinct from it, is the next portion, called the Thorax,—to which are attached the legs and wings. And then, thirdly,

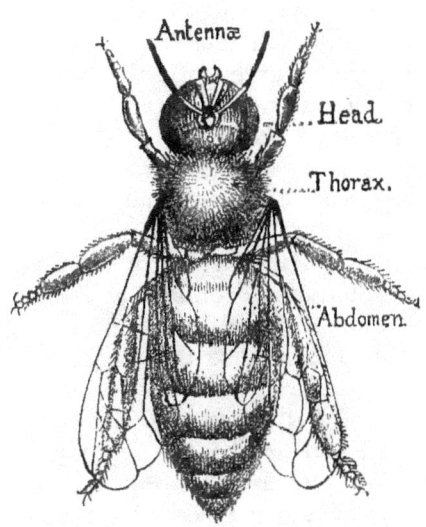

The Head, Thorax, Abdomen, of a Bee.

and very distinctly divided from the Thorax, there is what is called the Abdomen.

Some other points in which all true Insects are alike are the following: In the perfect state they all have six legs. They all have two antennæ, the peculiar thin long feelers which stand out from the head near the eyes. They also all breathe, not as animals with lungs, but through very small tubes, which run into all parts of the body, and have a multitude of very small openings through the side of the insect, but so very small that they can only be seen with a powerful microscope.

But then there are many thousands of different kinds of little creatures which have these points of resemblance, and which therefore belong to the great class of Insects. Consequently the next thing which has been done in the way of arrangement has been to divide this great class into smaller, but yet very large, divisions, or 'Orders,' as they are called.

This has been done by arranging them, chiefly, according to the number and character of their wings. I will not give you the names of all these orders of insects; it would only confuse you. But, as examples, all kinds of Beetles are put into one order, called Coleoptera, because the wings of all beetles have a hard peculiar sheath.

All kinds of Butterflies and Moths are put into another order, and are called Lepidoptera, because their wings are covered with a beautiful kind of scale-like dust, the scales being laid one over another like the tiles of a house.

And then we have the order of insects called Diptera, so called because they have only two wings, instead of four,—an order including the common fly, gnats, and many other such-like insects.

And then we have another great and important order, in which come Bees, Wasps, Ants, and many other insects, which go through a complete transformation. And as this order includes our bees, you must try and remember the long hard name by which it is called—Hymenoptera—so called because the wings (generally four in number) of all insects, belonging to the order, are of a thin kind of membrane.

Notice the wing of a bee, and you will see of what a thin and delicate, and yet strong, membrane it is composed.

But then, as this order of Hymenoptera is very large, and includes very different insects, although they all have the same membranous wings; the whole order (and it is the same with the other orders) is subdivided again into families,— each family being distinguished by its own peculiar character.

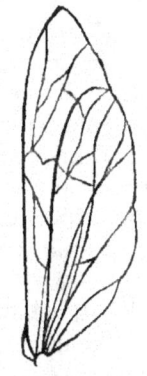

Wing of a Bee.

Thus we have the family of wasps, including all kinds of wasps; and the family of ants, including all the many kinds of ants; and then, amongst the others, the great Bee Family, called Apidæ, including all kinds of bees.

But then, again, this great family of bees—and now we must think only of this one family—includes so very many different kinds,—there are such numbers of bees, as was mentioned in a previous chapter,—that these again, according to certain points of resemblance, are put into divisions of their own, and are called genera.

Thus we have the genus Apis (a bee), and the genus Bombus (humble-bee), and many more.

And then of each of these genera there are many species. Thus of the genus Apis, with which we are specially concerned, our honey-bee is one species called the species 'Mellifica,' because of the honey that it gathers.

And then of this species of honey-bee there are many, so-called, varieties, such, for instance, as our English bees, and the Italian bees, and the Cyprian bees, and many more. These are varieties of the one species of honey-bee Mellifica,—a species which belongs more especially to Europe and the adjacent countries.

I think we had better now go over this rather hard lesson again. And if, for example's sake, we take a specimen of the Italian bees, of which there are so many in this country, we may think of it thus :—

First, it belongs to the insect 'class,' having, with other characters of a true insect, the three distinct parts—head, thorax, and abdomen.

Secondly, it belongs to the 'order' of insects called Hymenoptera, because of its four membranous wings.

Thirdly, it belongs to that 'family' of the Hymenoptera which is called Apidæ, or the Bee Family.

Fourthly, it belongs to the 'genus' Apis, as distinguished from the genus of humble-bees and others.

Fifthly, it belongs to that 'species' of the genus Apis, which is called Mellifica, or Honey-bee.

Sixth, and lastly, it belongs to that 'variety' of Mellifica, which is called 'Italian,' because of its Italian origin.

The following diagram will perhaps make it clearer :—

NATURAL HISTORY OF THE BEE. 65

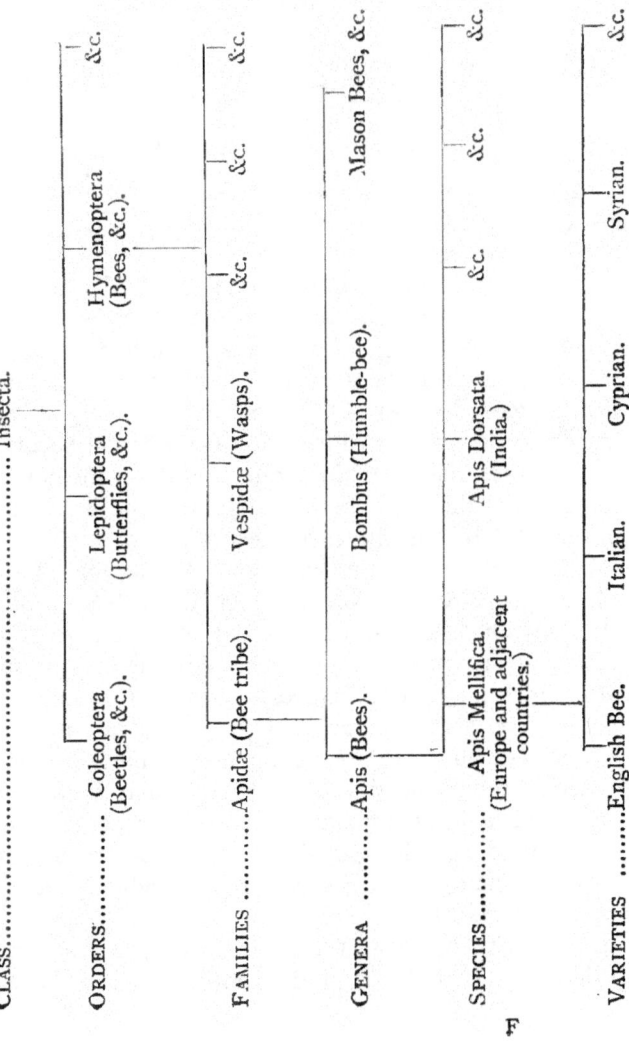

CHAPTER XVI.

THE TRANSFORMATION OF INSECTS.

THE next thing to think of is the way in which the bee is produced—born into the world ready for its busy, active life.

The bee—and it is the same with all insects—comes from a tiny egg laid by the mother insect. It is, however, an egg which greatly differs in many respects, besides its size, from the egg of a bird. Both eggs—the egg of the bird and the egg of the insect—contain that which, after a time, will become, as the case may be, the young bird or the young insect; but the process by which this end is reached is very different in the two cases.

You all know the process with the egg of the bird. Nurtured by the parents' warmth and care, the egg hatches, and produces the young bird; which, in most cases, is as helpless as any infant, although in some instances, as with the common chicken, it is able to run, and feed itself at once. In every case, however, the young one, immediately it is hatched, is without doubt a bird. It may be a poor, wretched-looking, unfledged little thing; but all the same, it is plainly a bird, and it goes through no further change. It only grows gradually to its perfect condition.

But with the egg of the insect the process is very different. It hatches, and produces, not an insect, but a grub or caterpillar,—a little creature as unlike as possible to the insect to which it will grow. In this condition it is called a larva—a name you must remember as we shall often use it. It does not, however, long remain a larva; for it has to go through two more changes before it becomes the perfect insect.

When first hatched the larva is very small, but it grows most rapidly, eating enormous quantities of food; so much so that the larvæ of some butterflies will consume in twenty-four hours double their own weight of food. Nourished by this abundant food, and grown to its full size, the next great change takes place, and the larva becomes what is called a Nymph or Pupa.

The process is very curious. The larva, in the first place, spins around itself a beautiful silken kind of web, called a cocoon. Within this covering the little creature—now called by its new name, pupa—begins to have, or rather to develope, its wings, legs, and other parts, gradually more and more becoming the perfect insect.

The time taken in this process varies greatly, according to the kind of insect. In some cases a very short, and in others a very long time is necessary. At last, however, the pupa state is over, and the day comes for the insect to issue forth into the world; and it breaks through its covering, and appears, to our astonishment, the perfect insect, now called the Imago—perhaps a butterfly, or beetle, or ant, or bee—but in all cases, with all its parts fully formed and full-grown, and itself able at the proper time to do its part towards bringing into the world other young ones like itself.

The process of these changes is called the Metamorphosis or Transformation of Insects. It is a most interesting subject, and full of wonders, of which I can now only just mention one, as a striking instance of the Divine Wisdom seen in nature,—namely, the way in which the food is provided for the grub in its larval state.

In some cases—as with bees, wasps, and ants—the food is provided not by the mother, but by other insects of the colony—in some instances by the mother herself. The little grub is thus fed and nursed, and taken infinite care of.

But in many cases, as with butterflies and moths, the eggs are simply left in some spot by the mother, who takes no more notice or care of them, but most probably herself dies very shortly afterwards. Now we might think that these eggs are left, without care, to chance, but it is not so; for the mother insect has selected just that place where, when the eggs are hatched, the young larvæ will find the kind of food they want. See how remarkably this is the case with the common white butterfly. You see it flying round some cabbage plants rather than the gay flowers. And why? It does not want the cabbage for itself; but it knows, taught by the marvellous instinct given to it, that the cabbage will afford the best possible kind of food for its young when hatched. Thus it goes to the cabbage, and there lays its eggs, covering them over with a thin case to keep them from the weather; and thus, when hatched, the young larvæ can immediately begin to eat, for their food is ready at hand.

As another instance of the same provision of nature, it may be mentioned that, in the case of a certain insect, it is necessary that its eggs should be carried by some means into the stomach of a horse. The insect itself cannot get there to lay its eggs, but it is managed in this way. The insect selects a spot for its eggs which it is likely the horse will lick,—it may be the horse's leg; and the horse thus unknowingly takes the eggs into its mouth, whence they pass to the stomach; and the object of the insect is accomplished.

CHAPTER XVII.

THE NATURAL HISTORY OF THE BEE—CONTINUED.

IN the case of the bee the process of transformation is as follows, and may be seen and traced in the following illustration.

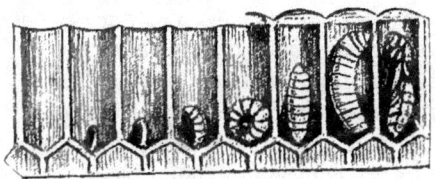

Egg and larvæ of the bee.

The mother or queen bee lays its small egg at the bottom of a cell. This is the young bee's cradle. There, the other bees (for the queen takes no further notice of it) surround it with a food, made of pollen and honey mixed into a sort of jelly. In three days the egg hatches, and there comes forth the tiny larva, which at once finds ready for it that kind of food which it needs.

Nourished with this food it grows rapidly, and, in the course of six days from the time of hatching,— or nine days from the time the egg was laid,— is full grown, and almost fills the cell, and is ready to begin the next, or pupa stage of insect life by spinning around itself the silken web of the cocoon. And, as it will now want no more food, but only to be left in perfect quiet, the bees who take care of it put a kind of cap or lid on the cell, and thus shut it in. They make this covering of very fine threads of wax and pollen beautifully woven together, but so contrived that the necessary air is admitted to the young one within.

In its sealed-up cell the pupa, following the rule of insect life, as before described, gradually developes into the likeness of a bee. Its legs and wings are formed. Its antennæ grow. Its mouth and other parts take their proper shape, and in twelve days more—or twenty-one from the time the egg was laid,—it is ready to come out of its prison-house a perfect, full-formed bee : and so cuts away the cover of its cell, and creeps forth, to be received with gladness by its companions who have taken such care of it in its helpless state.

> ' The full-form'd nymph clings to her close-seal'd tomb,
> Spins her own silky shroud, and courts the gloom.
> But, while within a seeming grave she lies,
> What wondrous changes in succession rise!
> Those tiny folds, which cas'd the slimy worm,
> Now thrown aside uncoils her length'ning form :
> Six radiant rings her shining shape invest,
> The hoary corslet glitters on her breast :
> With fearful joy she tries each salient wing,
> Shoots her slim trunk, and points her pigmy sting.'
>
> EVANS.

Such is the process with a worker bee, but with drones it is different. To obtain drones, some of the cells in the hive are made larger than the others, as explained before. In these the queen lays,—as she is able to do when necessary,—a different kind of egg. You would not, however, know it from the others. It looks just the same minute thing fastened to the bottom of the cell. But, when it hatches, it takes longer to become the full-grown larva; and then, when it is sealed up and spins its cocoon, the lid that covers the cell is made of that different shape, of which I spoke before, much higher and rounder, so that it is easily distinguished.

Out of this cell, and its pupa state, it does not come until some four days later than a worker, that is to say, on about the twenty-fifth day from the time the egg was laid. Then it comes forth, the great, sturdy bee, which makes so much noise and does so little work.

The way in which the queen is produced is one of the great marvels of the hive. In due course she passes through all the usual stages. First there is an egg, then a larva, then a pupa, and then, in due time, she becomes the perfect queen : but the remarkable thing is that the egg which produces this queen is not, so to speak, a queen egg, but an ordinary worker-egg which, under usual circumstances, would produce a worker-bee, but which, through the particular manner in which it is treated, and especially by the way the bees feed the young larva, becomes, not a worker but a queen.

What takes place is this : For some reason or another a new queen is required. Perhaps the old queen dies, or is too old to lay a sufficient number of eggs for the wants of the colony; or, perhaps, she is about to leave the hive with a swarm to find a fresh home. To provide for this want, the workers select one of the little eggs, lying at the bottom of a cell, or, possibly, a young larva, so long as it is not more than three days old. Then they enlarge the cell in which it is—very often treating several in the same way at the same time—by cutting away the cells around it ; and then, with other contrivances, build it out into that peculiar long shape, like an acorn as before described. (See illustration, p. 43.)

Into this large odd-shaped cell, containing the egg or very young larva, they put a quantity of jelly food, not of the ordinary kind, but jelly made in some peculiar way (it is called 'royal jelly'), the result of which is that the larva, when fed upon it, grows faster than it would if fed on ordinary jelly food, and, when five days old, is fit to be sealed up, and to go into its cocoon.

And now takes place the most marvellous change,

for in eight days more,—or about the sixteenth day after the egg was laid (instead of twenty-one in the case of a worker, and twenty-five in the case of a drone), it is ready to cut its way out, and to come forth, a beautiful young princess ; soon to become a perfect queen, and to begin to lay eggs.

What this food, or royal jelly, is, or whether there is anything else done or given, which turns the worker-egg into a queen-bee we do not know, but the fact is most extraordinary, for this queen-bee is, in many ways, a very different insect from the worker, which the very same egg would have produced, if it had not been treated to the large cell and the royal jelly.

As an illustration of this difference we may take the case of two dogs—the one a greyhound and the other a pug. If we put them side by side, the contrast is most striking. What can be more unlike,— the one with its slender legs, lithe body, beautiful pointed head, and quick, graceful movements, and the other with its short legs, square body, blunt nose and head, and ungainly movements? And yet there is not really so much difference between them, as between a queen and a worker-bee.

The difference between the dogs is in shape more than in anything else. They have mouth, and jaws, and teeth, in all points the same except shape. And it is the same with every part of their legs and bodies ; they have the same bones and muscles, and internal organs, however greatly they vary in size and appearance.

But in the case of the queen-bee, she not only has

a body differently shaped to that of the worker, but one that, in many respects, is actually different, wanting some things which the worker has, and having others which the worker has not.

Moreover, she is so made that her habits and instincts are quite different. And, more wonderful still, she will probably live two, three, or even four years or more, instead of only so many months; and be able, during her life to lay, an enormous number of eggs,—a million, or even more.

How marvellous is the change thus produced, so far as we know, by the wonderful food given to the larva! You see it is something far more wonderful than would be the feeding of the young puppy of a pug with some particular food, and by such a process of feeding, turning it into a greyhound.

CHAPTER XVIII.

THE STRUCTURE OF THE BEE ADAPTED TO ITS WANTS AND WORK.

IN a previous chapter something was said of the wonderful way in which bees are formed to accomplish the work they have to do. We will now pursue this subject a little further, and take one of the ordinary worker bees, which we have traced from the egg and its infancy to the perfect insect, and examine more closely some of its parts; and we shall see in it, I think, one example amongst countless others, how God, in His power, wisdom, and goodness, marvellously provides for all His creatures, and their wants.

To see this clearer, let me remind you, first of all, of one or two familiar examples. Such examples are on all sides. The very colour of animals is full of meaning. What, for instance, is more suitable than the brown colour of the partridge to hide it from view as it sits on the open field? On the other hand, what could give greater concealment than the white winter plumage of the ptarmigan on the snowy hills? The stoat, again, like the ptarmigan, is dark in summer-time, but often, in hard snowy winter, changes to white. Or, amongst fishes, what better to hide it from its enemies than the colour of the sole? Its under side is white, for this is not much seen, but its upper side is almost the exact colour of the sand on which it lies.

But, after all, nothing can better illustrate the great truth than the human body, and no part of it more so, perhaps, than your arm with its hand, fingers, and thumb, which is ever ready to obey your wishes, and with which you can do such different things as strike a heavy blow with a blacksmith's hammer, or pick up a little pin. Nothing can be more perfect than the arrangement of bone and joints and muscles and nerves. By no other possible arrangement could every part be so exactly fitted for its purpose.

We see it still more if we look at what answers to the arm and hand in many animals. They have bones, in some respects, similar to ours of the arm and hand, but then, in each case, they are just so altered as to make them exactly the best for the purpose of the animal.

Thus, these bones ' are recognised in the fin of the whale, in the paddle of the turtle and in the wing of the bird. We see the corresponding bones, perfectly suited to their purpose, in the paw of the lion or the bear.' *

The claws of the lion and the dog are other striking examples of the same thing. It is necessary for the dog to have claws rather to help it in running long distances, and to protect the foot, than to seize an enemy. And so this is just what it has. But the lion must not only have strong claws, but they must be kept sharp to seize and hold the prey. It would never do for its claws to be exposed like the dog's. They would soon get blunted and useless. And so by a beautiful contrivance the lion's claws are withdrawn into a sheath, and kept there till he springs on his prey, when at once they are brought into use, and strike deep into the flesh of the victim.

We are now prepared, I think, to find wonders of construction in the bee; and we shall not be disappointed. I will mention a few, but only the simplest. Some of the most striking, relating to the internal parts, you would not understand.

We will take, in the first place, and examine, one of the hind legs, of which here are illustrations when greatly magnified.

Now you will notice that it is divided into several portions, of which two are larger than the others, and of a peculiar flat shape, and if you look closely you

* *Bell on the Hand.*

will see that one is hollowed out, and that the hollow is made deeper by a fringe of hairs.

Nothing can be more perfectly constructed for what is required. It forms what is called the 'pollen-

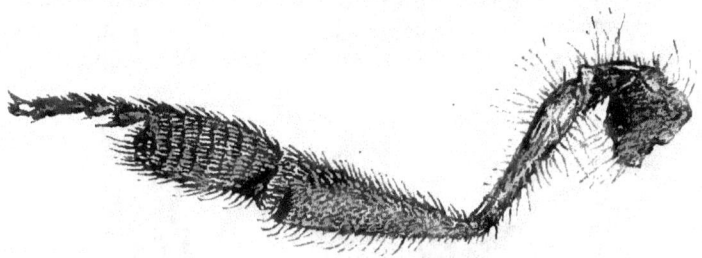

Fig. 1. Leg.

basket.' In this cavity, or pollen-basket, the bee places the fine pollen dust, which it gathers from the flowers,

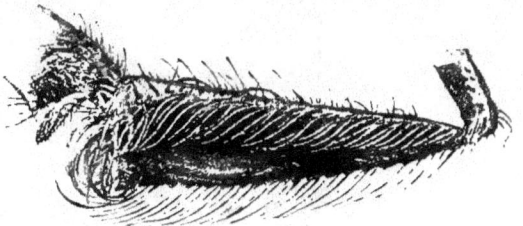

Fig. 2. Pollen-basket (reverse side of fig. 1).

working it into position by help of its other legs, and making it quite secure by the hairs which surround the little basket, some of which will be found buried in the pollen, and holding it very firm. It is wonderful how large pellets of this pollen the bee will in this way carry safely home, where it is removed, pushed off from the leg without difficulty, as the hairs point downwards.

At the end of the leg, or rather foot, there are two very small claws or hooks, which are most useful, and are adapted for many purposes. The bees, for instance, with their help will hang on to one another, until they make quite a rope of their bodies hanging,

Bee, showing Tongue.

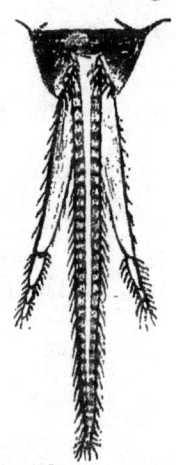

Tongue, highly magnified.

as they sometimes require, in the form of a festoon, from one part of the hive to another.

The legs are also covered more or less with hairs, which, like everything else, have purpose and are of great use. The bees use them as brushes to remove from their bodies the fine dust with which they are often quite covered, after visiting a flower.

In the next place, let us look at the tongue. Here are drawings of it.

You will wonder, I am sure, at its great length. It is almost as long as the whole body of the bee, but nevertheless is just the tongue the bee requires, for, when it goes to a flower, the honey is often very deep down, and otherwise would be quite out of reach.

The construction of the tongue itself is also very wonderful. It is made with a great number of joints, so that the bee can twist it about, like an elephant does its trunk; and, when it reaches the place of the honey in the flowers, can move it here and there and all round. And as it is covered with very small hairs, and the end of it is quite like a little brush, it sweeps up all the honey, which readily sticks to it, and which thus in a moment is drawn up into the mouth, from whence it passes into the honey-bag or stomach. And here it may be mentioned that this honey-bag is quite distinct from the true stomach, and simply a convenient place where honey can be stored till it is carried home.

In the next place let us look at the wings. I have said before that there are four, two on each side, one much larger than the other.

These wings, when not in use, are folded one over the other by the side of the bee, the larger wing on the top.

And now what could be more perfectly fitted for the purpose than the material of which the wing is made? You will notice, if you take a bee's wing and magnify it slightly, that it is so made as to be very thin and light, and yet very strong and tough. It is also, as you will see, strengthened with little ribs of stronger material.

I want, however, more especially to point out a most beautiful contrivance by which the wings are made, as it were, of double use. Of course, to fly fast, it is of great importance for an insect to have a large wing; but then a large wing, in the bee's case, would be very much in the way in the crowded hive,

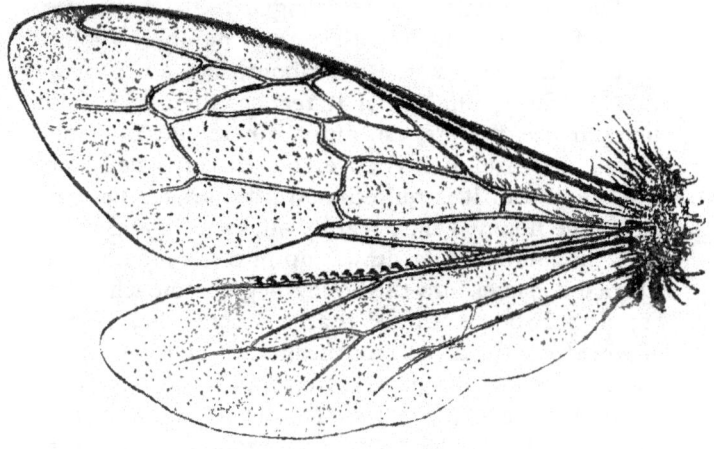

Wing—magnified—showing hooklets.

and when not in use. This difficulty is, however, partly got over by the bee having a second wing on each side, for, when both are spread together, there is a larger extent to resist the air, and so give power of progress.

But then, if this were all, as both wings beat the air together, the air, as we can easily understand would pass between them, and so half the power would be lost; just as it would be with the sail of a ship if it were torn down the middle. In such a condition it would indeed be of little use. Or again, if a lady's fan were divided into two portions it would take double the exertion, to get as much air from it, as if it were whole and in one piece.

Well, and so what do you think is done to help the bee in its flight? It is this. On the upper edge of the smaller wing there is contrived a row of very small hooks, and on the lower edge of the larger wing, just opposite these hooks, there is a sort of bar to which the hooks can fasten.

And then what happens is this. Directly the bee opens its wings to fly, the little hooks on the one wing catch hold of the little bar on the other, and in a moment the two wings are fastened together, and become almost like one large wing; but as soon as the bee stops the hooks are at once unfastened again, and the wings fold one over another, quite conveniently, out of the way.

Can anything be more strikingly beautiful than such a device? You will see the little hooks greatly magnified in the illustration.

CHAPTER XIX.

THE SAME SUBJECT CONTINUED.—THE STING.

THE next thing we will notice is the sting. Possibly you have already felt what a sting is like, and I hope you do not think it anything very dreadful. At all events, it is a curious fact that we can get so accustomed to stings that, although they may hurt us when we first begin to keep bees, they will hurt less and less, until at last they hurt so little that many bee-keepers care nothing at all about them.

But I want to speak of the sting itself, which is a very beautiful little instrument. You have of course seen a sting—the very fine little pointed dart which the bee shoots out and which pierces the flesh. This is usually called the sting, but it is not really so, for the sting itself is another still finer-pointed dart, which lies hidden in what you see almost as in a sheath. And this very fine inner dart,—which really consists of two, working side by side,—is barbed with sharp points, which prevent its being easily drawn back out of the wound. Connected with it is a very fine tube, which conveys a very minute drop of strong poison into the wound when the whole sting pierces the flesh.

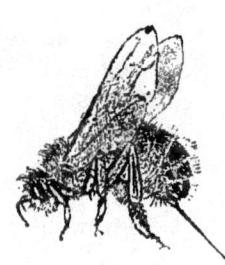

Bee and its Sting.

On account of the barbs, and the bee being unable to withdraw its sting from the wound, the whole sting, with its adjacent parts, is generally torn from the bee's body, and causes its death.

> 'With bite envenom'd they assail the foe,
> Fastening on his veins they shoot their darts
> Invisible, and in the wound expire.'
> VIRGIL.

When we consider the quantities of tempting food stored within the hive and the smallness of the little insect which has to defend the precious sweets against the covetousness of many enemies, we are surely led to marvel at the wisdom which has provided the insect with such a formidable weapon, and made it thus a match for even the larger animals.

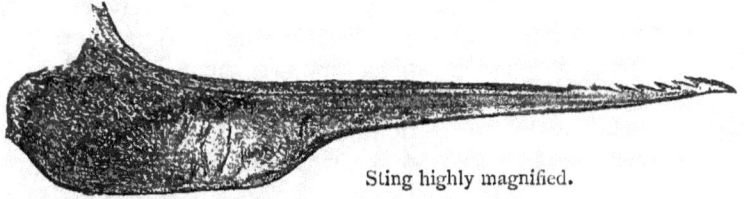

Sting highly magnified.

And here may be mentioned, as showing the exquisite perfection of the works of nature, that, as related by Bevan in his work on the Honey-bee :—

'Upon examining the edge of a very keen razor by the microscope it appears as broad as the back of a pretty thick knife, rough, uneven, and full of notches and furrows. And an exceedingly small needle being also examined, the point thereof appeared above a quarter of an inch in breadth, not round nor flat, but irregular and unequal, and the surface, though extremely smooth and bright to the naked eye, seemed full of ruggedness, holes, and scratches; in short, it resembled an iron bar out of a smith's forge. But the sting of a bee, viewed through the same instrument, showed everywhere a polish amazingly beautiful, without the least flaw, blemish, or inequality and ended in a point too fine to be discovered, yet this is only the case or sheath of an instrument still more exquisite.'

And now, passing by many wonderful things in the structure of the bee, such as the system by which it breathes, and the formation of the eye, and the internal organs, I will only say something of the antennæ.

All the uses of these most important organs we probably do not know, but, amongst other uses, they are certainly means by which the bees communicate one with another, and for this purpose are most exquisitely and delicately formed. When bees meet and, as their custom is, cross their antennæ, they undoubtedly speak to one another, whatever their language is.

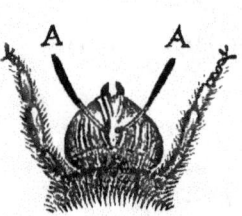

Head and Antennæ.

It is also evidently by the touch of the antennæ that they distinguish friends from enemies, and also by their use that they appear able to move, and work in the darkness of the hive just as easily as if they could see everything plainly.

A queen-bee that had lost its antennæ was observed by Huber to be itself as one that was lost in the hive—not to know its way about its own home, and only anxious, as soon as possible—quite contrary to the queen's usual instinct—to get out of the hive into the daylight.

One story will perhaps be sufficient to show their importance as means of communicating news, and that without them the bees cannot, as it were, talk to one another.

Into a hive full of bees a division was one day inserted, separating the whole colony into two portions one to the right and the other to the left. This division consisted, not of a solid board, but of two pieces of zinc side by side, and full of very small holes, too small for the bees to get through, but just sufficiently large for the bees to push their antennæ through. These two divisions, at first put in side by side and close together, were then separated an inch or two one from the other. The consequence was that, while the bees in the one half, where the queen happened to be at the time, were as quiet as usual, and went on working, the bees in the other half became in a very agitated state, as always is the case when their queen is removed.

But then, as the divisions were full of little holes, and not like thick board, why could not those bees, which had the queen on their side, tell the others that she was not really lost, but as well as ever? If they could have done this, all would have been well, and the agitation would have ceased, but this they could not do, and so the disturbance went on.

But now the two divisions were brought, gradually and slowly, nearer and nearer together, until at last they were so near that the bees could almost touch one another, but still the state of excitement on the one side continued. The bees on that side could not be satisfied as to the presence and welfare of their queen, but when the divisions were brought just a little nearer,—near enough for the bees on one side to touch with their antennæ the antennæ of the bees on the other side, then immediately all agitation ceased. The bees evidently at once knew their queen was safe, and this was quite sufficient, and so went to work again as usual and quite contentedly. It was a proof that it is mainly, if not entirely, by the antennæ that bees can communicate with one another.

CHAPTER XX.

STRUCTURE VARYING IN QUEEN, WORKER, AND DRONE.

WHILE reading the previous chapters respecting the construction of the bee, every part so exquisitely made for its purpose, you must not forget that what has been said applies mainly to worker bees. I mentioned this before, but call your attention to it again, because, when we look at queens and drones, we find many of those parts of which I spoke, such as the tongue, sting, and legs, strikingly altered, in their respective cases, to meet their special wants and work.

The queen, for instance, never leaves the hive to gather honey. It is not her work. Her duties are entirely at home, and so when we look at her tongue we find it unlike that of the workers, not so long, and not made to brush up the sweets from the flowers, but only fit to lap up honey already brought home, or to receive it from the other bees, who feed her when required.

So, again, the queen has no honey-bag in which to bring home honey from the flowers, and no little hollows or baskets on her hind-legs in which to carry the pollen, and no brush-like hairs on her other legs with which to remove the pollen dust from her body. All these are invaluable to the workers, but would be of no use to her, staying always in the hive.

Her sting also is different, for she has no occasion to use it against the common enemies of the hive. The workers are alone the fighting population.

And when we look at the drones we find the same adaptation of structure to the wants of the insect. We think of them as the idle ones, never going out to get honey, and doing no work at home; but indeed they could not gather the honey, or bring it home, or collect the pollen, even if they tried, for, like the queen, they do not possess a honey-gathering tongue. Neither have they honey-bag nor pollen-baskets. To collect food is not their work. And they could not fight, for they have no sting. Nevertheless you must not think they are useless. Indeed they are very necessary to the hive. They must be there if the hive is to prosper.

We can thus trace the workings of Divine Wisdom not only in the actual construction of every part of each kind of bee, but also in the way in which each is fitted for, and made to fill, its own little place in the community. One is queen, another worker, another drone, and to each one is given the means by which it can best fulfil its own duties, and be the most useful to the community at large.

And if so, we may be quite sure that the same wise and over-ruling Providence places each one of us in that position where, if we do our duties faithfully, we can be most useful; and that, instead of sometimes complaining of our lot in life, we shall do far better to try and make the best use of all the opportunities of work and usefulness that are given to us.

> 'How oft, when wandering far and erring long,
> Man might learn truth and virtue from the bee!'

Occasionally, as before mentioned, under certain circumstances we find a worker bee which, in the absence of a queen, tries to act the queen's part and to lay some eggs. But the consequences are most disastrous. The whole colony gets out of order: workers die, and only drones are born to take their place, and the colony soon altogether perishes. True example how each one should be content with the work of his own proper place, and not try to act the part of those in a different station of life; not to be the jackdaw assuming the peacock's feathers. It seems to tell us that we only do more harm than good if we try to do so.

CHAPTER XXI.

COMBS, AND THE FORM OF CELLS.

HAVING considered the bee itself—although there is a great deal more of the same subject which I hope you will learn some day—we will now look somewhat more closely at the house it builds for itself—how 'the singing masons' build their 'roofs of gold.'

I have spoken of this before, but only in general terms, describing how the combs are built of wax, with cells on each side, and so arranged that there is just space enough between the combs for the bees to work in. Now, I want to point out some more of the wonders of its construction, how the bees

> '. . . In firm phalanx ply their twinkling feet,
> Stretch out the ductile mass, and form the street,
> With many a crossway path and postern gate,
> That shorten to their range the spreading state.'
>
> EVANS.

And as we do this, I think we shall see it affording another instance of that marvellous instinct which guides the bee in all it does, and makes it the cleverest of architects and the best of builders.

We often talk of the wonders of engineering skill and man's ingenuity seen in countless inventions. We look, for instance, with wonder at our railroads and viaducts, and great bridges, and call them monuments of engineering skill.

There is, for instance, the marvellous great iron bridge across the Menai Straits, which hangs as a great iron tunnel suspended high up from rock to rock over the waters far below, and yet is so safe and strong that the heaviest railway trains are continually and with safety passing over it. No one can see it without admiration of the great skill with which it has been planned, and of the perfect workmanship shown in its construction. Everything is provided by countless and exact calculations to make it strong and secure. And it was just for the want of some of these calculations, and some consequent fault of construction, that on the night of December 28th, 1879, another great railway bridge, that over the Firth of Tay, in Scotland, failed to withstand the force of a great gale of wind, and in the darkness of the night, and when a train with a hundred passengers was passing over it, fell down, carrying with it into the deep waters below the whole train, not one passenger in which survived to tell the tale of the most frightful railway accident that ever happened.

Or look, again, at some of our great buildings, wonderfully contrived, skilfully constructed. If you go to Cambridge; there, in the magnificent chapel of King's College, you will see the whole wide space spanned by a roof of stone of enormous weight, which from below looks too flat to form an arch, and yet is so cunningly contrived and built with such skill that it stands perfectly secure.

But, after all, nothing equals the beauty and perfection of Nature's works seen all around us; and there is hardly a more striking instance of this than in the cell of the bee. It is absolutely perfection in every way, in plan and architecture, in material and strength, and in fitness for its purpose.

Take a piece of comb like that illustrated on the next page, and the first thing which we notice is the shape of the cells, that they are six-sided, or hexagons, all fitting in close together. And then, if it is a nice thin piece of clean comb, and we hold it up to the light, we shall see very plainly that the cells on one side do not correspond with the cells on the other, but just the reverse—the centre of any cell on one side corresponding with the spot where the sides of three cells on the other side meet together.

Then, if we cut away all the cells carefully and gradually, we shall find that we have, left in our hand, not a smooth piece of wax, such as would make the bottom of each cell quite flat, but a piece of thin wax, beautifully impressed with little diamond-shaped pieces put together, the bottom of each cell being

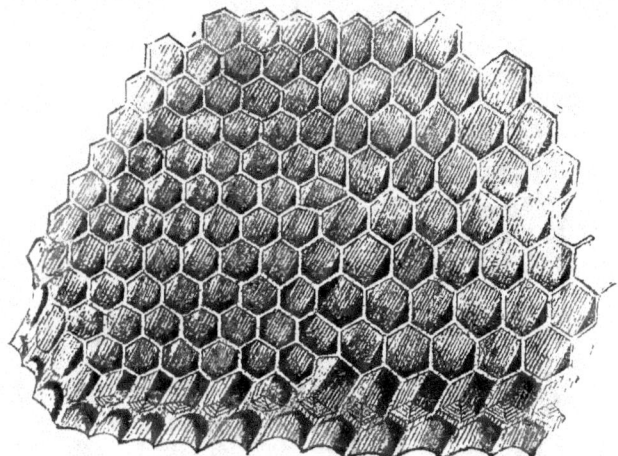
Comb—Worker and Drone.

formed of three such diamond-shaped pieces meeting in a point, as you see in the drawing below.

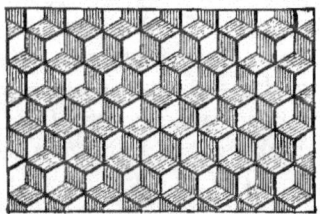
Comb Foundation.

And now, let us try and see how nothing can be more perfect than all this for the object in view.

First, then, we can easily understand that that form and make of cell will be best which economises to the greatest degree space, material, time and labour —all of which are very valuable to the bees—and also provides for the combs being the strongest possible, consistent with other requirements. They must also, at the same time, hold as much honey as possible, and be fitted, when required, for the rearing of the young bees. Here are a number of conditions to be fulfilled; and it is most interesting to see how marvellously the bees are led by their instinct to accomplish the task, and to get over the difficulties of the problem.

And first we will notice the hexagonal shape of the cell. Why is this the best? Why should it not be round? why not a square? why not an equilateral triangle?

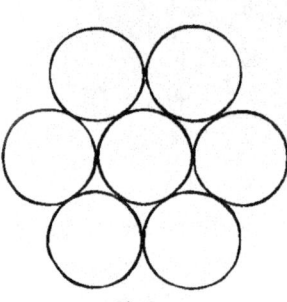
Circles.

Well, if the cells were round it would be better in one way, and, if the bees made single cells, standing out by themselves, I have no doubt they would make them round (some wild bees do so), for a round vessel can contain a greater quantity of fluid, in proportion to the extent of wall and material, than any other shape. If you were to take, for instance, the material of which a circular pint measure is made, you could not make it up into any other shape, having sides of the same thickness as before, so as to hold the pint as at first.

But then it would never do for the cells to be round, because what we gained in one way we should more than lose in another, for, if round, they would never fit together, and a great deal of space would be lost, and, not fitting, they would be very liable to break. Much heat also would be lost, a most important consideration to the bees. This you can see by the illustration.

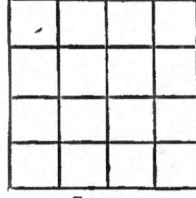
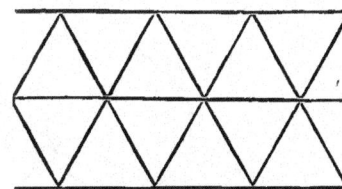

Squares. Equilateral Triangles.

But, if not round, would not a square shape do? Certainly not, for although square cells would very well fit side by side, all the corners (and it would be still more so with equilateral triangles) would be very awkward for the young bees, and, in making these corners and angles, a great deal of material would be wasted.

Well then, if the round shape would not do, but only because circular cells would not fit well side by side; and the square shape would not do, because of all the corners and waste of material; it follows (try and understand this) that the best shape for cells is that which is nearest the shape of the circle, and yet will allow the cells to lie close together. This shape is the hexagon, for although an octagon is more like a circle than a hexagon, a set of octagons would not fit together, only indeed a little better than a set of circles, but hexagons fit together perfectly.

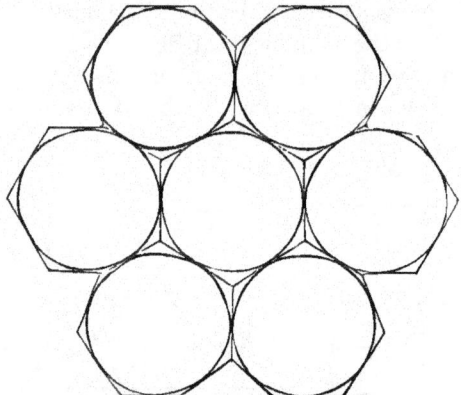

Economy of the Hexagonal Form.

Thus, in choosing the hexagon, the bees select the very best shape possible, that which enables the cells to hold the greatest quantity of honey with the least expense of material and waste of space.

Again, the bees want the combs to have great strength, and yet to have the walls of the cells very thin; the stronger the better and the thinner the better. But these two things seem contrary. If you want anything to be strong you generally make it thick. But the bees know better than you, and get over the difficulty wonderfully. While they make the walls of the cells as thin as the thinnest paper, yet by making them, in a most ingenious way, of two layers of wax joined together, they get a great deal of extra strength. Every cell has, as it were, double walls. And then again, by the way the cells fit together, and by the way in which they are arranged on each side of the comb, so as not to correspond, they get yet further strength without adding material.

Again, as the bees build a cell, gradually making it deeper and deeper, they always contrive to leave the edge, for a time, much thicker than the rest. The cell, in fact, has always a strong rim, which makes it firm to resist pressure and weight.

Then again, the arrangement of the little diamond-shaped pieces of wax at the bottom of the cells, of which I spoke before, is the most perfect possible. It is just that one plan which, more than any other, gives the greatest strength to the whole structure of both sides of the comb, and also good accommodation to the young bees. If there were the very slightest alteration of angle, so as to make the bottom of the cell either flatter or more pointed, the form of cell would not be so good for its purpose. This has been proved by mathematicians as plainly as that two and two make four. No architect or engineer, indeed, could possibly have planned all this better. But without any plan or calculation the bees know it all by instinct, and follow out this best way with the most astonishing exactness; and the result of their work is thus, as I have described, absolute perfection.

> ' These, with sharp sickle, or with sharper tooth,
> Pare each excrescence, and each angle smooth,
> Till now, in finished pride, two radiant rows
> Of snow-white cells one mutual base disclose,

> Six shining panels gird each polish'd round,
> The door's fine rim, with waxen fillet bound,
> While walls so thin, with sister walls combined,
> Weak in themselves, a sure dependence find.'
> EVANS.

There is also another difficulty which the bees get over wonderfully. If the cells were made horizontal, or at right angles to the middle partition, the honey would run out, almost as fast as put in, and so what the bees do is to make every cell slope a little inwards, and then, when the honey is put in, it is kept there, partly by what is called capillary attraction, and partly because, as they put in more and more, so much the more do they build up the entrance, until at last the cell is quite full.

Once more, there is another, and apparently serious difficulty which they meet with in comb-building,—but which they soon surmount most ingeniously,—arising from the drone-cells being larger than the worker. The width of four drone cells put together is one inch, which is the same as the width of five worker cells, measured in the same way. Consequently when drone-cells are built on by the side of worker cells, there is a difficulty in making them all fit together. Indeed it is impossible without contrivance and some alteration of shape.

How the bees manage you will best understand from the illustration of a piece of comb in a previous chapter, at page 91. There you see the two kinds of cells, the larger and smaller; and then how the bees make a few odd-shaped cells, which, being put in between the large and small cells, soon brings all back into proper shape and order. And the bees do all this in the dark!

'Is it credible,' says Langstroth, 'that these little insects can unite so many requisites in the construction of their cells, either by chance, or because they are profoundly versed in the most intricate mathematics? Are we not compelled to acknowledge that the mathematics by which they construct a shape so complicated, and yet the only one which can unite so many desirable requirements, must be referred to the Creator, and not to His puny creature? To an intelligent and candid mind, the smallest piece of honey-comb is a perfect demonstration that there is a great First Cause.'

> 'On books deep poring, ye pale sons of toil,
> Who waste in studious trance the midnight oil,
> Say, can ye emulate, with all your rules,
> Drawn or from Grecian or from Gothic schools,
> This artless frame? Instinct her simple guide,
> A heaven-taught insect baffles all your pride.
> Not all yon marshall'd orbs that ride so high,
> Proclaim more loud a present Deity
> Than the nice symmetry of these small cells,
> Where on each angle genuine science dwells.'
> EVANS.

CHAPTER XXII.

MORE ABOUT WHAT THE BEES DO.

WE pass on now to consider more fully than we have done before, some particulars of the *work of the bee* both at home and abroad. I have already said that every bee has its work, and works hard—works itself to death in a short time, but I want to point out a little more of the manner in which it works, and how it uses, and makes the most of, the various materials it gathers from the fields.

Of the queen I need not say much more. Her work is simply, as the honoured mother of the whole family, to lay the eggs which shall hatch into young bees to take the place of those lost by death, and thus keep up the full necessary strength of the colony, and furnish swarms for emigration. For this purpose she is made, and, beyond laying eggs, she does nothing,—never in any way taking care of her eggs after they are in the cells, but leaving all this to the workers.

But truly astonishing is the number of eggs the queen will lay,—as many as even 2000 or 3000 in the course of a day during the height of the honey season,— a very good day's work indeed! A queen has been seen to lay at the rate of six, or even eight eggs in a minute, putting each egg into its own cell; so that it is no exaggeration to say, as I mentioned before, that a queen, in the course of her life of three, four, or even five years, will lay more than a million of eggs.

The number of eggs, however, that she lays always greatly varies, not only with her age, but also according to the time of year and the weather. An old queen, as a rule, never lays so many eggs as a young one. She is generally at her best when from one to two years old. She will usually begin egg-laying in February, but instinct guides her not to begin before there is good promise of sufficient food to be had for the young larvæ when the eggs are hatched. She will, therefore, begin about the time when the early crocuses appear, as from these and some other early flowers the bees get a good supply of the food necessary for the infants. But if the weather is unfavourable, or the supply of food runs short, egg-laying is delayed; and, if already begun, at once, in a great measure ceases. And at such times, even if the queen wishes to lay, the workers will prevent her; they know the danger of having more young mouths than they can feed.

> 'The prescient female rears her tender brood
> In strict proportion to the hoarded food.'
> EVANS.

Aware of this instinct, Bee-keepers take advantage of it; and, when they want their queens to begin laying eggs rather earlier than they otherwise would, give them a little food—but only a little—day by day, which satisfies them that their little ones, if born, will not starve, and therefore that they need not fear to begin the great work of the year.

It is in May and June that the greatest number of eggs are laid. In September the queen generally, more or less, ceases to lay; although this mainly depends upon the weather, and the honey-giving plants of the locality, for she will sometimes lay eggs as late as November. Where there is heather, the breeding season is continued much longer than in other places. You see, thus, how in this as in other things, instinct guides the bee to do just the right thing at the right time.

But after all, perhaps, the most extraordinary fact about the queen is her power—as mentioned before—

of laying such eggs as will produce either drones or workers, just as required. When in the course of her egg-laying she comes to a drone-cell, she lays an egg which will produce a drone ; and when she lays an egg in one of the smaller cells, it is one that will produce a worker.

Of drones, also, we have not much more to say. You must remember, however, that they are the male bees of the hive, and that the queen finds a husband amongst them, but generally from amongst those of some other hive than her own. Beyond being necessary in this way, it is sometimes thought that the poor drone is quite useless in the hive. I feel sure that this is not the case, because, when we really and fully understand any production of Nature—even the smallest insect, or even the most minute part of any insect—we find some good reason for it ; and I am quite certain the drone in the hive is no exception to the rule. And although we do not, as yet, fully know all the good the drone does, or the use of the number of drones that we often find in a hive; I have no doubt that they serve one great purpose, and that is, to keep the interior of the hive nice and warm at a time when most of the other bees are out at work. The hive must be kept to a certain temperature, and always is so ; and if the drone can do nothing else, at all events its big, burly body gives out a great deal of heat.

When August arrives, however, the drones are no longer wanted for warmth or any other purpose, as the other bees stay much more at home. They are therefore in the way, and a very useless burden in the hive, eating a great deal of the food which is wanted for winter supply. The workers, therefore, now get rid of them,—drive them out of the hive, and leave them to starve.

> 'With terror wild,
> The father flies his unrelenting child.
> Far from the shelter of their native comb,
> From flow'r to flow'r the trembling outcasts roam,
> To wasps and feather'd foes an easy prey,
> Or pine, 'mid useless sweets, the ling'ring hours away.'
>   EVANS.

If the drones resist, the workers may be seen to seize them in the most determined manner, and without scruple to bite and gnaw their wings at the root, or wound them elsewhere; so that, when cast out, they cannot return, but are left helpless on the ground and soon perish from cold or wet. Resistance is useless for—

> 'All, with united force, combine to drive
> The lazy drones from the laborious hive.'
>   VIRGIL.

And is there cruelty in all this? Shall we blame the bees who thus destroy their companions whom they have reared with tender care? These are questions which we can hardly help asking; but, when we consider what striking proofs of wisdom we have on all sides, and how every creature of God is marvellously made and wonderfully provided for, and that nothing is done without good and sufficient reason, we cannot doubt but that there is good cause for the manner of death of the poor drone, as there is also for his apparent idleness.

CHAPTER XXIII.

THE SAME SUBJECT—CONTINUED.

HAVING considered the queen and the drone, we proceed now to think more particularly of the workers. I have spoken of their work before, in a general way; and, to make it all clear, I think it is well I should just remind you of what I have said on this subject in previous chapters.

Well, we thought a good deal of their industry, energy, patience, and cheerful work. I also described how they work early and late, out of doors, making even a hundred journeys in the day, if only the weather is fine, and the supply of food plentiful, and near at hand.

I also spoke of the way in which the bees, living together in a community, help one another, and work together, and thus, by united effort, produce the comb, the brood-nest, and the abundant stores of honey and pollen, and keep everything neat and in good order.

We saw how much of truth—even if somewhat of error—there is in Shakespeare's description of—

'The honey bees,
Creatures, that by a rule in nature, teach
The act of order to a peopled kingdom,
They have a king and officers of sorts:
Where some, like magistrates, correct at home;
Others, like merchants, venture trade abroad;
Others, like soldiers, armed in their stings,
Make boot upon the summer's velvet buds;

> Which pillage they with merry march bring home
> To the tent royal of their emperor:
> Who, buried in his majesty, surveys
> The singing masons building roofs of gold,
> The civil citizens kneading up the honey,
> The poor mechanic porters crowding in
> Their heavy burdens at his narrow gate,
> The sad-eyed justice, with his surly hum,
> Delivering o'er to executors pale
> The lazy yawning drone.'
> *King Henry V.*, Act i., Sc. 2.

I also, in a previous chapter, endeavoured to explain some of the reasons of the form in which they build the comb—how marvellously they make it just in that shape, and in that way, which gives the greatest strength and capacity, with the least material and space.

What more is there to say of their work? Well, a great deal more might be said, and I must pass over many things. I will, however, mention a few facts of interest—first of all, respecting their work in the fields and gardens, and then of their work in the hive.

First, then, of their work abroad.

> 'The winter banish'd and the heavens reveal'd.
> In summer light, they range the woods, the lawns,
> They sip the purple flowers, they skim the streams;
> Soon urged by strange emotions of delight
> To cherish nest and young.'
> VIRGIL (by Kennedy).

It is a question often asked, 'How far will bees go from their hives in order to find, and bring home, the honey?' I dare say you will like to know, and I am sure that what I have to say will surprise you, and make you feel, more than ever, what wonderful little insects our friends are.

Generally speaking, animals, birds, and insects do not go very far in order to obtain food for their young. In order to supply their own wants, and when they have no home with young, they will, as we all know, go far and wide; and many birds will migrate from one country to another; but when they have young—as the bees have in their hives—their journeys are limited. Rooks and pigeons will go some distance; so will foxes, amongst animals; but I imagine there is hardly any animal, bird, or insect that will go so far as the little bee.

The way in which this has been found out has been by marking bees in a particular way, and then going to some distant favourite place, and there finding the marked bees.

'A gentleman, wishing to test this fact, dusted with fine flour his bees as they emerged from a hive. Then, driving to a heath five miles distant, which he knew to be much frequented by the insects, he soon found many of those which he had sprinkled at home.'[*]

But even more wonderful than this, cases have been known of bees actually going seven miles from home on the same errand. At the same time, however, we may say that two or three miles is, perhaps, quite the limit within which the bees can collect honey with much profit. The stores collected from a greater distance cannot repay the extra labour and time expended.

[*] Harris.

But, on the other hand, the actual time occupied in any journey is not long. It is only, when many journeys have to be taken, that it is of much moment, for the swiftness with which bees fly is very astonishing. They very soon cover a mile of ground. We see them dart from their hives, and in a moment they are out of sight ; but, great as this pace is, I have no doubt, when out of sight, and 'the steam is up,' they go faster still. We gather some idea of this from what we have seen when travelling in a fast train with the carriage windows open. A wasp, or bee, attracted by some sweets within, will fly in and out of the windows, apparently as easily as if the train were at rest. On the other hand, a partridge, frightened by the passing train, and flying along the line, will hardly keep pace with the carriage in which you are seated.

Another remarkable fact connected with the bees' work is, that when in search of honey and pollen, they do not go from one kind of flower to another, but always keep to the same kind during any one journey. Whatever the kind of flower they begin with, they go on with, until ready to return home. They do not, for instance, go from mignonette to sweet-pea, although both may be growing in the same border ; but if they begin with mignonette, they go on with it, and so with the sweet-pea.

One would have thought that they would go to the flower which came most conveniently in their way, without making any selection ; but such a mixture would never do. So, if you examine a little pellet of pollen when brought home, you may find it deep yellow or light, or it may be red or brown; but you will not find these colours mixed. It will be all of one colour, coming from one kind of flower. You will hear, later on, why this is, and that it is one of those wise provisions of the Ruler of all, which gives us what is beautiful and profitable in our fields and gardens. At present, however, I only want you to remember the fact.

CHAPTER XXIV.

MORE ABOUT HONEY, POLLEN, AND PROPOLIS.

AND now, what is honey? Is it something made, or only gathered? You know that it comes from the sweet liquid sometimes called 'nectar,' which is produced, or, as it is termed, secreted, with considerable rapidity by the flowers, especially when the weather is warm and sunny—so much so, that a bee may in such weather go very frequently in a day to the same flower, and take all away, and yet, when it comes back, find more ready for it.

If, however, it were possible, and we ourselves were to collect all this same sweet liquid, it would hardly be what we call honey, and it would soon become acid. But when collected by the bee, it undergoes some slight change in the honey-bag, and then, when it is put into the cells, the bees are very careful not to seal it up at once. They leave it for a time so that all the watery liquid in the honey may pass away or evaporate. It then becomes thick, and will keep good for a great length of time. And thus, although, as I have said, the bees do not make it, they do something more than merely collect it.

In flavour, it varies very much according to the source from whence it comes. The very best honey is gathered from the white clover, although some people think that no honey is to be compared with that which is gathered from the heaths.

When the bee goes from home, to gather pollen, it often undesignedly collects it over its whole body; for in many flowers the pollen is like the finest dust, which is shaken off in clouds as soon as the flower is touched. The bee then has to get it off its body, and on to its pollen legs. This it does by means, as before described, of its other brush-like legs; but it is sometimes so covered that you will see it return to its hive like a little miller, when the other bees come to its help and remove it all.

When, however, in the process of honey-gathering, the pollen sticks to its tongue, we may well ask how it gets it off, and on to its pollen legs? This might well seem difficult, but, like every difficulty, it is provided for. On the fore-legs of the bees there is a very curious little notch. You will see it in the illustration, which is that of a portion of the leg magnified. It is thus described by Root:—

'There is a little blade, as it were, at B, that opens and shuts; and the bee, when its tongue is well loaded, just puts it into the grooved or fluted cavity, then shuts down B, and gives its tongue a wipe so quickly that it leaves conjurors all far in the shade.' This little notch is also used in the same way for cleaning the antennæ. How marvellous this contrivance! You see again how everything has an object and use.

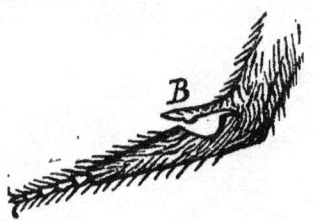

Leg with notch, magnified.

But then it has to get this sticky pollen from its fore-legs into the pollen-baskets. How does it manage this? Well, between the pollen-gathering legs and the pollen-basket legs are another pair, and these play a very important part in the operation. With the tongue, fore-leg, and middle leg, the bee pads up the pollen and honey until there is quite a wad of it, and then, with a very quick motion, almost too quick to be seen, carries this little cake, scarcely so large as the head of a small pin, between the middle and fore-leg, back to the pollen-basket. When in place, it is firmly pressed, and then neatly patted down with the middle leg, and so is ready to be carried home.*

The propolis is carried home by the bees in the same way as pollen. The bees gather it chiefly from the sticky buds of certain trees, such as the chestnut,

* From Root.

fir, and poplar; and also from the gum which oozes out through the bark of these and other trees.

> 'With merry hum the willow's copse they scale,
> The fir's dark pyramid, or poplar pale,
> Scoop from the alder's leaf its oozy flood,
> Or strip the chestnut's resin-coated bud.'
>
> EVANS.

The bees use it, as you have been told before, for several purposes, but chiefly for filling up all cracks and chinks, which otherwise would let in cold air. But, at the same time, they are quite ready to put it to other uses as occasion offers.

Here are two curious instances, showing how cunningly and ingeniously they contrive to meet difficulties. A snail once crept into a hive. What could the bees do with it? They could not sting it through its shell. They could not drag or drive it out. What they did was to surround the edge of the shell with propolis, and so to fasten it down tight to the floor of the hive. A little was sufficient, and all air was excluded, and the snail was, as it were, buried in its own shell.

On another occasion a slug entered a hive. This the bees soon stung to death. But then, how were they to remove it? And, if left, the smell of its decay would be unbearable. This apparent puzzle the bees soon solved, for they at once covered the whole body with a coating of propolis, which made it quite as harmless as if it had been buried.

> 'For soon in fearless ire, their wonder lost,
> Spring fiercely from the comb th' indignant host,

> Lay the pierced monster breathless on the ground,
> And clap in joy, their victor pinions round.
> While all in vain concurrent numbers strive,
> To heave the slime-girt giant from the hive,—
> Sure not alone by force instinctive sway'd,
> But blest with reason's soul-directing aid,
> Alike in man or bee, they haste to pour,
> Thick-hardening as it falls, the flaky shower;
> Embalm'd in shroud of glue the mummy lies,
> No worms invade, no foul miasmas rise.'
>
> EVANS.

Who of us, indeed, with all our reasoning powers, could have thought of a better plan?

CHAPTER XXV.

WAX, AND HOW THE BEES MAKE IT.

OUR next chapter must be about another, and very important material of the hive, of which at present we have said but little. We have thought of the comb and its form, now we must consider the wax of which it is made. The questions which suggest themselves are these: What is wax? How is wax made? At one time it was commonly supposed that wax was made of the little pellets of pollen which the bees were seen to take into the hive. Now, however, we know better, and that, although pollen may have something to do with it indirectly, wax is really made of honey, and honey alone, by a most curious and elaborate process. I must not try to explain it all to you, but in order to understand a little of the process you must first look at the under side of the abdomen of a bee, and there, if the bee is occupied in comb-building, you will see some very small flakes or scales of wax sticking to it in several places, which places are often called 'wax-pockets.' These little flakes of wax are produced from the honey in the honey-bag, which undergoes a certain course of preparation within the bee, and then is secreted, and appears, not as honey, but as wax. Generally speaking, these little bits of wax can only be produced when the bees are in a great heat; and thus, when they require to make wax, they first of all have their honey-bags full, and then have a way of hanging together in what looks like a solid cluster, but which really consists, so to speak, of a great many ropes of bees clinging to one another. In this curious position they remain perfectly quiet, and great heat is produced.

Then, after a time, the little wax-scales make their appearance, and these, when duly formed, the bee carries away to the place where it is wanted, and where other bees fashion it into the required shape. It is not, however, quite fit for use as it comes from the wax-pockets, and, before using it, the bees mix it with a kind of saliva, and knead it up with their jaws.

It thus takes a great deal of time and trouble, and a great many bees, to make a little wax; so much so that it is a fact that the bees consume as much as twenty pounds of honey to make one pound of wax, so that it is in every way a very expensive material, and it is of great importance to the bees that they should make as little of it as possible.

It used to be thought that the wax-makers were a special set of bees by themselves, but this is not the case. All the workers, more or less, take their turn, except the very young bees.

And now of these young bees, these children of the hive, I must say something that you must try and remember. It is this—that these young ones do not leave the hive for two or three weeks after they are born, except at times for short flights to play and take exercise like children. All this time they remain at home to get strength, and (shall we say?) be taught their duties. But, although they stay at home, they are by no means idle. Do not think this for a moment. Indeed, they have most important work to do, and they do it like useful children.

First of all, the task is given to them of looking after and nursing the young grubs in the brood nest. For these they prepare the food, and put it in the cells; and then when the proper time comes, seal the cells over, doing everything that is necessary. These young bees are often called the nurses, and very good nurses they are.

They are also in great measure, although not always, the comb-builders, taking the wax from the wax-makers, and fashioning it into the proper shape. They also do much other work, storing away into the cells the honey and pollen brought in by the other bees.

They are, indeed, very useful young bees, very helpful to the mother in the care of her little ones, and although not old enough to go out into the fields

The Cottage Hive of Helpful Children.
From a drawing by Charles Jenyns.

—the wide world—yet quite ready to do anything at home which is within their power; and in this set an example to children who, even when quite young, should be cheerfully ready, as far as possible, to assist their mother, always seeking to be helpful children.

CHAPTER XXVI.

NIGHT-WORK AND VENTILATION.

AND now is it not wonderful to think that a great deal of all this work goes on at night,—more indeed at night, when all the bees are at home, than in the day, when many are absent? Except during winter the bees are always hard at work; they improve not only the 'shining hour,' but the darkest hour. They never leave for another day, or even hour, what can be done at once. If any repairs are needed they well know by their wonderful instinct how true it is that 'a stitch in time saves nine.'

Virgil makes the mistake of saying that they sleep at night:

'. . . . When eve at length
Admonishes to quit the balmy field,
Home to refresh their weariness they come;
Awhile about the doors and avenues
Thronging with drowsy hum, till in their beds
Couch'd for the night, a silence o'er them creeps,
And all their busy life is lull'd to rest.'
VIRGIL (by Kennedy).

Very much more might be said on all this subject

of the bees' work, what they do and what they make, but I will only mention one other thing. It is another kind of work, and very hard work, although it is all done while the bees stand still in one place!

Go to a hive in summer time, and you will see, at the entrance, several bees (and there are many more inside doing the same thing) standing with their heads toward the hive's entrance, and keeping up the most rapid movement with their wings; and so intent are they on their work that they give no heed to anything else, although many bees, going into the hive, may knock against them, and almost go over them. On they go with their work until quite tired out, when others take their place. What is it all for?

Well, it is for what is called ventilation, in order to blow, as with a fan, a quantity of fresh pure air, from the outside, into the hive to take the place of that which has become bad and unwholesome, owing to the number of bees and the confined space. As the good air is forced in, the bad air is forced out. It is the same with the hive as with our own rooms. These, as you know,—or ought to know, —must have their windows regularly opened, or, at all events, fresh air let in by some means, for, if not, they become most unwholesome, particularly if many people are in them. Nothing is of more importance. It is absolutely necessary to health, and you must always remember it. And so it is with the bees, and they know it; and, as they cannot open windows, they adopt this ingenious plan of blowing in the fresh air by their wings; and so thoroughly well does it answer the purpose that, however hot the weather, they always manage, unless there is disease in the hive, to keep the air in a pure state.

Besides the bees who thus ventilate the hive you will see others also at the entrance, acting as guards, watching for any enemy or strange bee, but in a moment recognising their friends by a touch of their antennæ, and letting them pass. See, however, a fly or a wasp come near, and out they rush at once, ready to fight boldly, if necessary, even to the death.

> '. . . . Some are bid
> To keep strict sentry at the outer gate,
> And take their turns of watching cloud and rain.'
> VIRGIL.

CHAPTER XXVII.

THE DIVISION OF LABOUR IN THE HIVE.

IN the next place, we see in all this varied work a striking example of the importance and results of division of labour. You have all read stories about this, how not even a little pin is made without a great number of people—men, women, boys, and girls having had part in it. Or you have read how, when a house is built, although only a few hands are seen to work upon it, thousands have really done something towards it, in preparation of materials, in bringing them by rail and ship to the spot, in making the tools, and so on.

> 'So works the honey-bee.'

As each day comes round each bee has its special work. Some gather honey, some pollen, some propolis, and, of those at home, some are ventilating, some guarding the entrance, and others are attending the queen, or are wax-making, or storing the honey and pollen, or nursing and feeding the young, and so on:

> 'Some o'er the public magazine preside,
> And some are sent new forage to provide;
> These drudge the fields abroad, and those at home
> Lay deep foundation for the labour'd comb,
> With dew, Narcissus leaves, and clammy gum,
> To pitch the waxen flooring some contrive,
> Some nurse the future nation of the hive;
> Sweet honey some condense, some purge the grout,
> The rest in cells apart the liquid nectar shut.'
> VIRGIL (by Dryden).

Thus:

> 'Each morning sees some work begun,
> Each evening sees its close.'

And by division of labour, as well as by hard work, they bring about their great results.

And then another thing in all this work of the bee, which we cannot fail to notice with admiration, is the great importance which they attach to little things, teaching us that it is by sticking to our work and attending to little things that we shall best succeed in anything that we have to do. Just look at the bee's care and attention to the smallest things —to do the smallest things in the best way. And observe again—as mentioned before—how they are never wasteful. It is indeed but very little that any one single bee can do. According to careful calculation, any one bee does not collect more than a teaspoonful of honey in a season. And yet see what is brought about by all thus working together, and all doing their little, and putting their little stores together. See the full hive as the result. And even the full hive is not all, for they will sometimes make 100 lbs., or even more, of honey—over and above all they store in the body of the hive—which the bee-keeper may take as the reward of his care and trouble.

But after all, if we only take notice, all nature around us is full of the same great lesson—how much can be done by little workers and care of little things.

One of the most curious and wonderful examples has been pointed out and explained by Mr. Darwin,—the great naturalist, and perhaps the closest and most ingenious observer of nature who ever lived. We think the little worms the most insignificant of creatures; but he has shown that what the little worms have done, and now are doing, is most astonishing. The worm throws up its tiny 'worm-cast,' and we think nothing of it. It is the most trifling thing possible; but in the course of ages these little morsels heaped together have been the means of changing in appearance large tracts of land.

It is, perhaps, more wonderful still to look at the lofty chalk cliffs of our sea-shore, and to know that they were formed in the course of countless ages by the work of some of the smallest of insects—insects so minute as only to be seen by a microscope.

And, yet again, we see the same in the mighty

work of the little coral insects, which, in countless numbers through countless ages, raise from the depths of the sea in tropical climates, islands and reefs of coral rock; which in many places form harbours of shelter for great ships, and, in others, most dangerous hidden rocks, upon which many a good ship has been wrecked.

Well, with these examples before you—and especially that of our friends, the bees—learn the value, and learn to make the most of, little things. Let me remind you of some good old sayings: 'Waste not, want not;' 'A pin a-day is a groat a-year;' 'Take care of the pence, the pounds will take care of themselves.' Yes, 'take care of the pence.' Put your pence into a Savings' Bank. There is a Savings' Bank at nearly every post-office, where you may do this. Or perhaps you have a Penny Bank in your parish. I could tell you many stories of such a bank in a country village, where many a child by taking care of pence soon became possessed of pounds; but now I can only say that you may look at the hive as a great savings' bank. The bees, with care and labour, put in their little gatherings; and the result is plenty for themselves, and plenty for us as well.

'Little drops of water,	'Little deeds of kindness,
Little grains of sand,	Little words of love,
Make the mighty ocean,	Make our earth an Eden
And the beauteous land.	Like the heaven above.'

CHAPTER XXVIII.

MORE ABOUT THE OBSERVATION OF BEES.

IN a former chapter I spoke of the importance of keeping our eyes open, and that, if we do so, we shall see wonders all around us; and I spoke of Huber, the great observer of bees, and how he discovered many things although blind. But I suppose we should never have known many of the facts of which I have told you, without the help of what are called 'Observatory Hives.' Such a hive is made, as you see in the illustration, with glass slides or large windows, and of such little depth between back and front that it will not hold two combs side by side. There is, however, just room for one comb between the two glass sides or windows; and the consequence is that every bee in the hive can be seen, either on one side or the other. The glass sides have wooden shutters; but the bees soon get accustomed to having them open, and go on working away as usual while you are looking at them closely.

Observatory Hive.

Thus through the glass you will easily and plainly see the queen, as she walks over the combs laying her eggs, and surrounded by her attendants; and you will see all the care of the nurse-bees—how they feed the larvæ, and how the comb is made, and the cells filled with honey and pollen. And as these observatory hives are generally kept in a room, and have a way for the bees to go out and come in through a little hole in the wall, there is no difficulty about observing everything without danger of being stung.

How to manage one of these hives you will perhaps learn at a future day from other books. These hives are a comparatively modern invention, but even Huber had something of the kind, which he called a 'leaf-hive.' It was made like a book with three or four leaves, each so-called leaf containing one little comb, the bees getting into the leaves by a common entrance at, what we may call, the back of the book. Although far inferior to the modern observatory hives it was another proof of his great skill and ingenuity.

We will now conclude this part of our book with one more example of what can be done by observation.

Sir John Lubbock, who, as I described before, made such interesting experiments as to the daily work of bees, and who has made many others respecting their hearing, smelling, and affection for one another, was anxious to determine how far bees, as they fly from flower to flower, can distinguish one colour from another; and he contrived to discover it in the following ingenious manner.

First of all, he got a bee from one of his hives to

come to some honey, which he put upon a small piece of glass, placed upon some coloured paper. After the bee, which he marked with paint, had become well accustomed to go backwards and forwards, carrying some of the honey to its hive—and while it was away—he arranged near the glass first one and then several other pieces of glass, each with honey, but each with a different-coloured piece of paper underneath. Thus, when the bee came back from time to time, there were pieces of glass with honey looking different to its first original piece— perhaps blue, or red, or yellow. But, although all might be tempting, the bee knew its own colour, and went to its old place.

But now further to test it,—while it was away—the paper under its own piece of glass was removed, and made to exchange place with another bit of paper; so that in the old place, although glass and honey were the same, they appeared of a different colour. And now what did the bee do? Soon it came back, and was going straight to its old place, but saw at once that things were altered; and so stopped and hovered for a moment, but soon caught sight of its own colour, and went straight to it. In other words, colour was, to a certain extent, its guide to the food. This experiment, after it had been tried again and again, and in various ways, was conclusive that bees do know something of colour; and therefore can distinguish one flower from another by colour.

By a series of further experiments he found out that if bees have any preference to one colour more than another, it is to blue.

CHAPTER XXIX.

INTRODUCTION TO BEE-KEEPING.

IF you have read the former part of this book with attention, you now feel, I hope, some interest in the subject of bees, and see that they are indeed marvellous little insects, deserving of all care and attention. As a consequence, I hope you feel that you would like to keep bees, and see for yourself some of the wonderful things of which I have been speaking. And I am quite sure that, if you only have a suitable place in which to keep them, and, chief of all, if you have got, as it is termed, 'a head on your shoulders,' and a kind heart to love all God's creatures—'all things both great and small'—and to treat them well, you may thus keep them, and find enjoyment in the pursuit, and also get some profit in the shape of money for the savings' bank. Boys and girls of thirteen or fourteen years old may very well keep and manage one or two hives.

But how can you make even such a start as this? Well, I will tell you. You must begin bee-keeping by keeping together your pence and sixpences—by saving up with care—until you have got together perhaps ten shillings, or a little more, with which to buy a stock or swarm of bees in a straw hive. This will be a small beginning, but it is best to begin in a little way. There is a proverb which says, 'Who goes slowly goes long, and goes far.' And again it has been wisely said, 'To know how to wait is the great secret of success.' A great many people fail in bee-keeping because they try to begin with everything at once; and do things on a large scale with modern inventions, before they have had any experience of practical management, or have tried their hands at some of the very simplest things.

It is with bee-keeping as with every other pursuit, you cannot get up the ladder of success all at once. You must begin with the first round, and get higher step by step, using, first of all, simple means, with care and industry. 'Fortune favours industry.' Smiles, in *Self Help*, has well said, 'The greatest results in life are usually attained by simple means and the exercise of ordinary qualities. The great highroad of human welfare lies along the old highway of steadfast welldoing, and they who are the most persistent, and work in the truest spirit, will invariably be the most successful. Fortune has often been blamed for her blindness, but Fortune is not so blind as men are. Those who look into practical life will find that Fortune is usually on the side of the industrious. Success treads on the heels of every right effort. Nor are the qualities necessary to ensure success at all extraordinary. They may, for the most part, be summed up in these two—common sense and perseverance.'

Very thoroughly does this apply to bee-keeping. You will succeed if you exercise 'common sense' and 'perseverance.' First of all, then, make up your mind to take trouble in the matter. Remember, that if there is anything you can do fairly well without trouble or difficulty, you will generally be able to do it much better by giving it some thought. Determine then that you will succeed with bee-keeping, and that, at all events, you will not fail through negligence.

It is very sad to see the poor bees in some gardens, uncared for and neglected, put away in some damp, dismal corner. They are thus often left to themselves, to live or die; and yet people wonder why others get honey and profit, and they get none. I remember once being asked by the lady of a large country-house to examine some hives in the garden. They were not successful. They made no honey. It must be a bad country, or a bad year. Such things were said. But what did I find? I remember well one miserable straw skep, rotten and broken down, with a large hole rotted through the top, through which one could see the combs and the poor bees at work—a hole letting out all heat, and letting in the rain. It was a melancholy sight! Think whether you could live in such a house, almost tumbling down, with the windows gone, and the roof partly off, and all damp and cold! Poor bees! What could they do in such circumstances? It was a satisfaction to be able to save their lives.

It is a good old saying that, 'if a thing is worth doing at all, it is worth doing well.' And I am sure it is so with bee-keeping; so that I hope, before you get even a single hive, you will resolve to manage your bees in the very best way you can. Try to excel in the management. It was said of the great Lord Brougham that 'such was his love of excellence, that if his station in life had been only that of a shoeblack, he would never have rested satisfied until he had become the best shoeblack in England.' And such efforts to excel will not only give the satisfaction of success, but, in the case of bee-keeping, will best bring actual profit.

Good management always pays. We see example of this every day in every condition of life. Some people, by management, seem to make a shilling go as far as two shillings in the hands of others. Some people, for want of management, are always behindhand in everything, and always in trouble in consequence. For a garden, for instance, good management, as well as good labour, is necessary. A great deal of work may be done, but unless it is well-directed work, or, in other words, unless there is management, much of it will be thrown away. It is the same with a farm. If it is worth while to farm at all, it is worth while to farm well; and the better the land is farmed, the better will it pay. In short, management and labour must go together in order to bring success.

Virgil gives us a good example of all this, describing his visit to an old gardener of his day, who, in all he did, fully carried out the great principle, and by labour and management took first place both with his garden and his bees :—

> ' For once do I remember to have seen
> . . . . . .
> An old Corycian gardener, who possest
> A few scant acres of forsaken ground,
> For pasture or for ploughing all too poor,
> Ungenial for the vine ; yet here he rais'd
> His vegetable fare, verbenas, lilies,

> Esculent poppies in the brake he sowed,
> Rich as a king in happiness ; and home
> Returning late at eve, his frugal board
> With unbought dainties cover'd : first was he
> To cull the vernal rose, the autumn fruit ;
> And when a wintery frost was even yet
> Splitting the rock and fettering the stream,
> That old man shore the soft acanthine leaf,
> Chiding the zephyr and the spring's delay.
> Therefore his hives the first with offspring teem'd
> And swarms abundant; soonest would the combs
> Their foaming juices to his pressure yield :
> The pine, the linden flourish'd best with him ;
> And every blossom that with beauty clothed
> His orchards to autumnal ripeness grew.'
> VIRGIL (by Kennedy).

Of course in bee-keeping, as in other things, there may be unavoidable failures. There are often bad seasons for bees, and there are summers cold and wet, when but little honey can be gathered, but perseverance, as I have said, will win the day at last. It is said that George Stephenson, the great engineer, when addressing young men, was accustomed to sum up his best advice to them in these words, 'Do as I have done—persevere.'

CHAPTER XXX.

FIRST PRINCIPLES OF BEE-KEEPING.

I will now tell you a little about bee-keeping, but you must quite understand that this book is not intended as a guide-book. Such a book, you will have to get ;

and I especially recommend one called *The British Bee-keepers' Guide-book*, by T. W. Cowan, or another called *Modern Bee-keeping*, published at 6*d.*, by the British Bee-Keepers' Association.

All that I can now tell you about bee-keeping will be the chief principles on which you must act. I want you so to understand these first principles that you never do anything simply because a book tells you, but rather because from what it tells you—you understand why you are to do it, and why it is the best way.

And now talking of first principles, the first great rule to be observed is, of course, Never to kill the bees for the sake of the honey. The old-fashioned way of murder in the sulphur-pit must entirely be done away with; you know, from what has been said before, what a cruel, foolish, and improvident way it is, and I need say no more on the subject.

But the next great principle I must explain more fully. It is, Always to take care to have a great number of bees in every hive—to keep the colonies strong. This is called 'the golden rule' of bee-keeping. You will learn that this can be done, and how best to do it, at a future time from guide-books. Now I only ask you to remember it as a great principle of successful bee-keeping, that the greater number of bees there are in the hive, the better the work goes on. And this is the case, not only because there are more workers to bring in honey, and 'many hands make light work,' but because a better heat is kept up, and the bees work with greater spirit. It is also a curious fact, as I will explain at a future time, that in winter a great number of bees in a hive will eat less food than a smaller number.

I can also explain the advantage of a number of bees in another way. Suppose I have a hive that contains 40,000 bees; and of these, we will say, 30,000 go out to gather honey, and 10,000 stay at home to keep it warm. And then—to compare with this strong hive—I have two others, each containing just half the number of bees, namely, 20,000. In each of these two hives the same heat must be kept up, as in the first hive, and, to effect this, the same number of bees—namely, 10,000—must remain at home, and so only 10,000 can go out from each hive to gather honey; that is to say,—adding these gatherers together,—we have only 20,000 gatherers from both hives, whereas we had 30,000 from the one hive at first; so that we have actually 10,000 more gatherers from the one strong hive than we have from the two weak ones put together. Always remember, then, 'the golden rule.'

And now if you are going to 'keep bees' you very likely ask, What kind of bees shall I keep? Are they to be Italians, or the common bees? And, what kind of hive am I to get? If you ask my advice, I would say that you had better not trouble yourself with such questions at first. You will learn, after a time, that Italians are the best and most profitable, and you will learn a great deal about hives, but never mind all this at present. Probably you want to begin without spending much money, and if so, your best plan will be, as I have said, to buy a stock of common bees in a straw skep, and wait a year, and see what you can do with it.

K

If you can get the constant advice of some near neighbour who understands the modern hive, you may begin with one; but, generally speaking, it will be better for the first year to stick to the straw skep; and if you manage well, you will by that time have gained valuable experience, and also a little money with which to buy a better hive, and to begin more thoroughly. Only take care, when you buy the stock, that you get a good, strong, and healthy one; and one that is not more than a year or two old, and one that has a young queen. Ask some bee-keeper of experience to help you in your purchase. You will also do well to get it early in the year, even if you give a little more for it; for then you will soon get a swarm from it, and so almost begin with two hives instead of one.

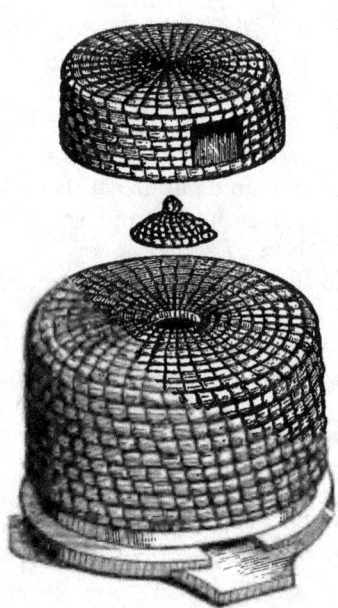

Straw Hive and Super.

In the next place, be a little particular about the shape of the straw skep. It ought generally to be one of medium size—although this may vary according to locality; and have a good hole at the top, closed with a cap or cork. And, above all, let it have a flat top, as in the illustration, because you will want to place upon it, what is called a super, in which the bees will make honey for you to take, and this super will not stand well if the top is round. And then by some means you must manage to keep the hive dry; placing over it some kind of covering, so that no rain can reach it to make it damp. And, when you choose a place for it to stand, see that it is well sheltered from cold winds, and in a situation where, as far as possible, the early sun will shine upon it. A guide-book will give you other directions, but these are the chief things to remember.

Formerly the straw skep was nearly the only kind of hive used, and many bee-keepers even now, prefer them to others. And certainly such hives have their advantages. They do not require so much care or trouble as other hives; and bees thrive very well in them for a time, for the straw is a very good material to keep the bees warm in winter, and at the right temperature in summer. And although they are called old-fashioned, they may easily be kept without there being any necessity to kill the bees in the old-fashioned, cruel way. And very good honey may be obtained from them, although not nearly the quantity which we get from the more modern hives.

But at the same time they have their disadvantages, and I am only advising you to get such a hive just to begin with, and that you may get accustomed to the bees; and also that you may, when you want it, get a swarm to put into a better hive. The disadvantages, indeed, are so many and so serious tha

I hope you will not be content without soon having something better.

To give you some idea of these disadvantages: suppose something goes wrong with the bees—some of their enemies get inside, or the bees are ill, as sometimes is the case;—you know that it is so, that things are wrong; but what can you do? There the bees are, safely shut up in a hive, where you cannot either see or help them. There are things you could do, and remedies you could apply, which would soon put all straight again, but you are helpless to do anything.

Think how it would be if you yourself were in similar circumstances. We will suppose that you are ill, and the doctor sent for. He comes, and feels your pulse, and asks you many questions, and so prescribes his medicine. But what could the doctor do if, when he came, he had to stand outside the house, and not even see you through a window? And of what use would be all his medicine if, when brought, it had to be set down outside the house, and there was no one to bring it in? You see, of course, the absurdity of the whole thing. And yet this is much the case with the bees in a straw hive when things go wrong. We cannot get at them, either to see what is amiss, or to give them any remedy.

And then there are frequently other occasions when many things can be done to assist the bees, and to make them into thriving colonies, if only we are able to see into the hives, and to handle the separate combs.

And now, keeping all this in view we see not only

the objections to straw skeps, but also what are the chief points to be observed in the construction of any good hive. It must, we see, be so constructed that we can easily and thoroughly examine it in every part, and, if necessary, see every bee and all that is going on within—what is right and what is wrong.

CHAPTER XXXI.

THE FRAME-HIVE AND THE PRINCIPLES OF ITS CONSTRUCTION.

In the modern hive, all that I have spoken of in the last chapter, as most necessary, can be done perfectly and with ease. This is the Moveable Comb or Frame-Hive. It is constructed in a hundred different ways, but in all there is the great principle of the moveable frame. In a general way it may be said that the principal part of the interior of the hive is simply a warm, dry box, of a certain size, and made very exactly to that size. It may or may not stand upon legs; and, instead of an ordinary flat top, it has a roof, like a house, so made that it can be lifted off without difficulty.

But we are now chiefly concerned with the frames, which are shaped as shown in the illustration, and of which there are a number—at least ten—with which the body of the hive is filled. All these frames can easily be taken out of the hive. They have, you

THE FRAME HIVE.

see, shoulders (these shoulders are made of various patterns) which rest on ledges running along the sides of the hive. They are made, also, just a little smaller than the inside of the hive, so that when they are in their places they hang quite loose and free, as seen in the illustration on the next page. You will thus easily understand that if we can only get the bees to make their combs in these frames, and exactly straight and true, we shall have obtained what we wanted, and be able to lift them out, one by one, just as we require, and see every part of the hive and every bee.

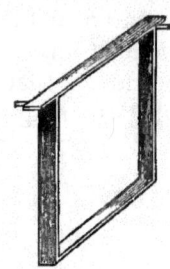
Frame.

If, however, the bees do not build their combs straight and true in every frame, but crossways or crooked, we are no better off than without them. So that this is the first essential thing with a frame-hive —to get the frames filled properly with comb. The hive may be very beautifully made, but all is useless unless the bees build their combs exactly straight and true in all the frames.

To get the bees to do this might seem very difficult. Formerly it would have been thought out of the question. It is, however, really the easiest thing possible, by means of sheets of wax, called Comb Foundation, ingeniously made of the required size, and in a manner most helpful to the bees.

This comb foundation is made by dipping a flat, smooth piece of wood, which has first been wetted, into melted wax one or more times (like a tallow

THE FRAME HIVE. 135

dip), according to the thickness required. The wax which adheres to each side of the wood is then easily peeled off in sheets. Afterwards, these sheets

Hive showing Frames in position.

of wax are pressed by a machine, and run through rollers, which have, cut upon them and all over them, the exact resemblances of the beginning of comb-cells. And so, when the sheet is passed between the rollers and is finished, it has received on all parts, and on both sides, impressions just like the commencement of cells.

These sheets—at least, if for use within the hive—are also made of just that thickness of wax which gives the bees sufficient material with which to lengthen out, and to finish off, the cells thus begun for them, so saving them the time and trouble of making any more wax. For use in supers a very much

Frame Empty.   Frame with Foundation.

thinner kind is made, for this is used more as a guide for the bees, to show them which way we wish them to build their combs, rather than as a help to them in wax-making.

These sheets of comb-foundation are fixed without much difficulty, perfectly straight within the frames. When all are thus filled they are put into their places within the hive, and well covered over with proper material—generally layers of flannel or carpet. And then, when the bees of a swarm are put into the hive, they are so delighted to find such good provision for them, and almost half their work of comb-building already done, that they at once and without hesitation set to work to make and finish their combs out of these sheets in the frames. And the result is that every comb, when made, is in a frame by itself, and true and straight as we wanted. And thus, without difficulty, the first great and essential thing is obtained. Each frame is independent of the others, and can be lifted out with comb and bees upon it whenever wanted.

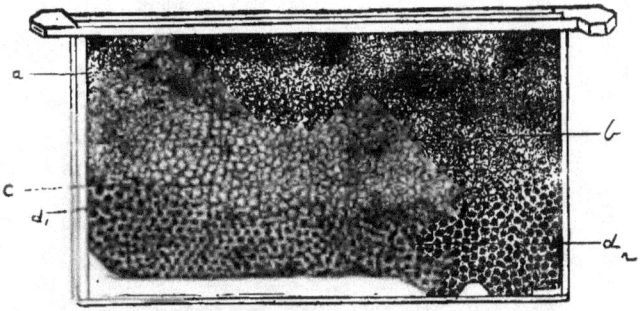

Frame filled with Comb.
(*a.*) Sealed honey stores.   (*c.*) Sealed drone comb.
(*b.*) Sealed worker-brood comb.   (d_1.) Unsealed drone comb.
(d_2.) Unsealed worker comb.

All frames—whatever the sort of hive—are now made of the same size, namely, 14 inches long by $8\frac{1}{2}$ inches deep. This is called the standard size. They are also made just of the proper width to hold a comb; and so contrived at the shoulder that, when in the hive, they are exactly at equal distances from one another, and just at that distance which the bees choose when making comb for themselves.

And now going back to the hive itself, I will point out a few more of the principal conditions to be fulfilled before it can be called a good hive. You will find full particulars and many such conditions in the best guide-books; but the mention of three or four will be sufficient for our present purpose. Of these the first is that the hive must be well and strongly made of good-seasoned wood. It may be very rough, but it must be strong, and must not warp or crack. In the next place, the inside of the hive, where the frames hang suspended, must be exactly of the right depth and width. It should hold ten frames at least, and it would be well if there is room for several more; for, if not filled up with frames, the vacant space is cut off from the bees by a stout division-board, and is always useful when examining the hive.

But whatever the number of frames, the width and depth of the hive inside must be true to measure; so

Section of a Hive with Frames.

made that when a frame is suspended within, there is a space of half an inch between the bottom of the frame and the floor of the hive, and a quarter of an inch between the sides of the frame and the sides of the hive. And there is great reason for this exactness. If the space at the sides is greater than a quarter of an inch the bees will build comb in it, and so fix the frames; and, if less than a quarter of an inch, they are unable themselves to get round the frames as they require, and so they fill up the space with propolis, and fix everything tight, which is worse still. The space below the frames the bees require as they come rushing in with their loads, to carry them to all parts of the hive.

In the next place, whatever the description of hive, it must be warm in winter, and not too hot in summer.

This is best provided for by its having double walls with, if possible, a space between. Good hives, however, may be made with single walls if only the wood is of sufficient thickness. But, whatever the walls, you must always remember that it cannot be a good hive unless, as I say, it is warm in cold weather, and the bees inside are protected in summer from the scorching heat of the sun.

Again, our hive must be very dry. This is quite essential for the bees. They cannot live in a damp house. If therefore the hive stands out in the open, it must be well painted, and must have a good roof, well made, to throw off the rain. And then this roof —which must either be hinged to the hive, or made separately, so that it can be taken off—must have plenty of room inside. You will hear the reason of this presently, but remember it as a necessity—a good high roof with plenty of room inside.

And now I have really told you the chief points of a good hive. There are other things of importance, such as the size and construction of the entrance, and how it should be sheltered ; and, within the hive, how the frames should be kept true in their proper position, and how thick should be the flannel covering over them ; but of all these things I shall leave a guide-book, or some bee-keeping friend, to tell you.

With these few simple rules to be observed, I think it very possible that at some future time, if you have intelligence, and a ready hand to use a few simple tools, you may wish to make a hive for yourself. Only if so, it will be best for you, in the first instance, to purchase or to obtain the loan of one as a model. You will hardly succeed without this, although there will be no occasion for your hive to have the polish and finish of first-rate workmanship. Your home-made hive, indeed, may be a very rough one, but all the same very serviceable, if only you copy your model and abide by the first principles I have mentioned, and do not substitute fancies of your own. In any case, however, you will do well to buy the frames, which can only properly be cut by machinery, and cost a mere trifle.

If in the construction of your hive you can plan and contrive with old material, and manage to use odds and ends of wood, without the expense of buying new from the carpenter, your interest in your hive, when completed, will, I think, be all the greater. And it is always wonderful how the exercise of ingenuity will get over many difficulties in such things, and find some way of adapting to the end in view the most trivial things possible.

Here is an account how some of the greatest men, distinguished in after life in science and art, began in the most humble way as boys, contriving to work out their schemes, and to practise their art with the most odd things possible :—

'A burnt stick and a barn-door served Wilkie (the great painter) in lieu of pencil and canvas; Bewick (artist and engraver) first practised drawing on the cottage walls of his native village, which he covered with his sketches in chalk; and Benjamin West (afterwards President of the Royal Academy) made his first brushes out of the cat's tail. Ferguson laid himself down in the fields at night in a blanket, and made a map of the heavenly bodies by means of a thread with small beads on it, stretched between his eye and the stars. Franklin first robbed the thunder-cloud of its lightning by means of a kite made with two cross sticks and a silk handkerchief. Watt made his first model of the condensing steam-engine out of an old syringe. Gifford worked his first problem in mathematics, when a cobbler's apprentice, upon small scraps of leather which he beat smooth for the purpose; whilst Rittenhouse, the astronomer, first calculated eclipses on his plough-handle !' *

CHAPTER XXXII.

SOME ADVANTAGES OF THE FRAME-HIVE.

WE proceed now to think of some of the advantages of the frame-hive. Some of these are plain enough. For instance, you are able, as I have said before, at

* Smiles.

any time thoroughly to examine such a hive by taking out the frames one by one, and thus see all that is going on within. Again, if you have more than one such hive, it is always possible and generally easy, to make one hive help another, as occasion requires. If one hive is weak and another strong, the bee-keeper will take from the strong hive some of the frames containing brood or honey, as may be needed, and give them to the weak one.

Then, when winter comes, the bee-keeper will take some of the frames away, and confine the bees to a smaller space, and thus make the most of that heat which is so necessary to the welfare of the hive. These frames he can return to them in the early spring when food is needed.

Again, a very important part of bee-keeping is to take care that the queen in every hive is young and healthy. In a straw skep the bee-keeper cannot find or see the queen, but with a frame-hive he can easily do so; and when she is too old to be useful, she can be removed, and another queen given in her place. There are many ways of doing this, as also of obtaining young queens, but I cannot now explain the process. Here, again, a guide-book will help you.

I may, however, just mention how a new queen is generally given to a hive. It is an operation very interesting, but requiring care. If the new queen were merely put into the hive instead of the old one, there would be little chance of her life. The bees, faithful and loyal to their old sovereign, would kill the intruder. They sometimes do this by stinging her, but more generally by encasing her, as it is termed,—clinging round her, making her the centre of a ball of bees, and so suffocating or squeezing her to death.

Consequently when the bee-keeper wishes to give a new queen to a hive, he first removes the old one, and gives the bees a little time to mourn her loss. Then, when they are beginning to prepare to make a fresh one for themselves, he puts his new queen into a little wire cage, like one of these here illustrated,

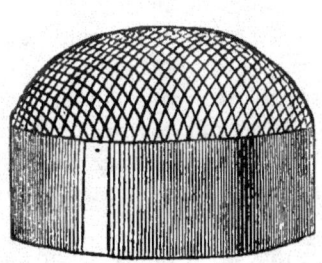 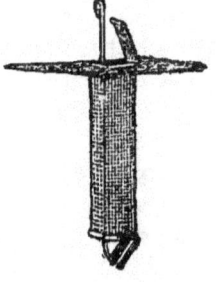

Queen Cages.

and then puts cage and queen carefully into the hive. The bees at first may wish to destroy her, but are prevented by the wire covering of the cage. After a while they get used to her, talking to her through the wires, and are ready to adopt her as their own queen. As soon as this is the case, the bee-keeper lets her out, and she is welcomed, and the hive once again prospers with a young and active queen and mother at its head. But one of the greatest advantages of the frame-hive is that the frames, when the combs are full of honey, can be taken out, and by means of a machine called an Extractor, which will be presently described, be emptied of their sweets without destroying the combs, and then put back again for the bees to refill. This cannot, of course, be done with the combs in a skep. A very great quantity of honey can be taken in this way, for the bees, being spared all the trouble of making fresh comb, very soon fill the empty cells, and the bee-keeper can again and again, during the season, take away a good supply of honey out of the same combs.

Another great advantage of the frame-hive is that the bee-keeper can, without much difficulty, at the proper time of year make an increase of his colonies, just as he thinks fit. He can, as a good guide-book will explain, make one hive into two; or, which is an excellent plan, he can make three out of two, or four out of three. He is indeed the bee-master as well as the bee-keeper, and the bees are his most willing and industrious little servants.

Then, again, with the frame-hive, many enemies can be destroyed, diseases cured, new frames given when required, the number of drones regulated, and many other things done, most helpful to the bees; so that, as I have said, you must not always be content with the straw skep, although, to begin with, it is a good hive.

But now, doubtless, the question has been suggested to your mind. Yes, but when you talk of lifting out frames, and finding the queen, and doing all that has been described, will the bees permit you? Will they not so attack and sting you as to compel you to give in? No, not so; for they patiently submit. Some people think there is a great mystery

ADVANTAGES OF A FRAME-HIVE.

in this, and that bee-keepers have some secret charm. There is, however, nothing of the sort, unless it is the charm of gentleness, kindness, and a knowledge of the bees' habits; and nothing can compensate for this. But then to avoid being stung, you can wear a veil; and, if necessary, gloves, although these latter are but seldom needed, and indeed do much to aggravate the bees.

Bee-veil.

The greatest help, however, is obtained from smoke, a very little of which, puffed into the hive, will generally very soon quiet the bees, and make them almost as harmless as flies. The effect of the smoke upon the bees is very curious. In the first place, it frightens them,

Smoker in Use.

and the result of their alarm is that they instantly run to the honey-cells, and fill themselves with the sweets. It is supposed that instinct teaches them thus to prepare, if necessary, to leave their home, carrying with them as much store as possible. But having thus filled themselves with honey, they generally become in the best possible temper, and seem to put away their stings; just like a bad-tempered man, who is always in better humour after a good dinner than when hungry.

Some bee-keepers, however, hardly use any smoke, and some use a preparation of carbolic acid, a very little of which applied with a feather to the tops of the frames answers almost as well. A spray diffuser with sweetened water is also very useful at times.

CHAPTER XXXIII.

SUPER HONEY AND THE EXTRACTOR.

I MUST now describe a very important part of bee-keeping, namely, how to obtain the honey, which the bees are ready, if managed well, to store away for us, either in large boxes or glasses, or, far better still, in those beautiful little cases which are called sections. But before I describe the process, let us think of the state of things within the hive which leads the bees thus to prepare and fill them.

As summer advances the hive becomes more and more populous. Young bees are hatching out daily, and all the frames, not wanted for the brood-nest, are more or less filled with honey, so that there is really neither room nor work in the hive for all the bees, The consequence is that an emigration on a large scale must take place.

The same thing happens in our own land. England becomes over-populated, and so thousands leave the old country, and go across the seas to find new homes, and fresh land to cultivate, in countries where there is abundant room for all. We hear thus of thousands going to America, New Zealand, and Australia.

In the case of the bees, when this kind of emigration must take place, and they feel the time coming near, they begin to make preparation by, first of all, taking steps to provide a new queen as formerly described. When she is nearly ready to come out of her royal cell, scouts go out to find a favourable home for the emigrants. It may be in an old tree, or in the roof of a house. I have often known bees at such times come down a chimney, black as sweeps, into a room. These were the scouts looking for a new home, and examining the chimney-pot; and, being unable to return, on account of getting covered with soot, they fell down into the room. All things being now ready, the bees on some fine morning, if left to themselves, would issue forth as a natural swarm.

But it often happens that the bee-keeper does not want any increase in the number of his stocks. And so, before this swarming takes place, indeed as soon as the hives become full of bees, he says, ' No, I do not intend you to make a swarm. You must all remain at home, and make honey for my use.' And when the bees reply, 'We cannot do so, for we have no room in which to store it,' the bee-keeper still says, 'No; I cannot let you go, but I will give you room. I will give you a large super, or a glass, or more probably a number of the little sections; and, as soon as you have filled these, I will give you more. You shall never stand still for want of room; you shall always have plenty.'

To carry out this purpose, the bee-keeper first of all prepares his super or his sections by fixing in them small pieces of comb foundation, made very thin for this special purpose. This secures

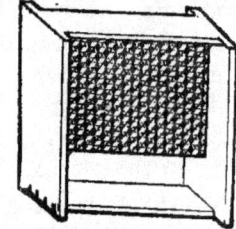

Section with Foundation.

that the bees build the comb, true and straight, in the proper shape. Then, in the case of sections, he puts a number of them into a wooden frame called a rack, which holds them all together.

Sections in Rack.

In the next place he takes off from the frames in the hive the quilts and covering, and puts on the whole case of sections instead, covering all over with plenty of flannel. And then, lastly, he puts on the roof, which, as I have said before, must have plenty of room inside. You see this done in the following illustration.

The bees now, of course, have easy admission into all these sections, and generally will begin work in them without delay if only the sections are kept

Hive with Sections in Position.

perfectly warm, and the weather is favourable. And very soon some will be finished, filled with the purest comb and the best of honey.

As soon as some are filled, and the cells sealed over, the bee-keeper takes them away, substituting at the same time an equal number of empty sections. Doubtless the bees are extremely astonished to find empty sections instead of full ones. They must think, that it seems like an endless task to try and get them full. It must seem to them like the task of Sisyphus in the old story, who had to roll to the top of a hill a large stone, which had no sooner reached the summit than it fell back, and the labour had to be recommenced. But like good bees—giving us an excellent example—they make the best of it, and go on honey gathering, humming their pleasant song, content to work not only for themselves but for others.

> 'The more their strength calamity hath drained,
> The more will they exert them to repair
> The nation's falling state, their garners fill,
> And re-construct their masonry of flowers.'
>
> VIRGIL (by Kennedy).

But when the bee-keeper wishes to obtain the greatest possible quantity of honey, he does not depend entirely on supers, whether sectional or of any other kind, but uses to a great extent the machine called an extractor, to which allusion has been already made. More can be obtained by its use than in any other way; but then it is not honey in the comb, and has not that beautiful and inviting appearance which belongs to well-filled sections.

The extractor—which is also very useful for other purposes—is a very ingenious machine. There are several forms of it, all more or less made after the pattern of those perfected by Mr. Cowan, and in all there is the same principle. The frames, when taken from the hive, have, first of all, the caps of the cells removed with a sharp knife. They are then put into a kind of wire cage (fig. 2), which, being placed in the machine are made to whirl round and round with great rapidity. The effect of this is, that the honey is thrown out of the cells against the sides of the extractor, and running down, can then be drawn away from a tap at the bottom.

The force which brings this about, and throws out

the honey, is called 'centrifugal force.' I may illustrate its action in this way. If you tie a piece of

Fig. 1.

Fig. 2.

string to a stone, and then, while holding the end of the string in your hand, swing the stone round and round, the stone is always, by action of this same force, trying to fly off. And the faster you swing it round the more the stone makes effort to get free; and if you let it go, it flies a long way. It is much the same with the honey, as the comb which contains it is swung round in the extractor. The difference is that, as the capping is off the cells, the honey can get free—instead of being confined, like the stone by the string—and so is thrown out.

CHAPTER XXXIV.
MORE ABOUT SWARMS.

I HAVE previously described the issue of a swarm, and the state of things within the hive which leads to it—that it is a forced emigration on a large scale. I have also told you how the bee-keeper often obtains artificial instead of natural swarms when, and as he thinks fit. And I need not add much on this subject, except that it must always be remembered that a good swarm, whether natural or artificial, must be an early one, according to the old saying :—

> 'A swarm in May is worth a load of hay,
> A swarm in June is worth a silver spoon ;
> A swarm in July is not worth a fly.'

It is not true, however, that a July swarm is so worthless, for bees are always useful, and such a late swarm, if not returned to its own hive, can be given to some weak hive in want of bees.

Of second swarms, or 'casts,' as they are called, I must say something. When a first swarm departs it leaves behind it a comparatively empty hive, but one that soon will be full again. It also leaves behind it a queen-cell, out of which in a day or two will issue a new sovereign to take the place of the old one who left with the swarm. There is always one such queen-cell—and generally several—at such swarming-time, each containing a young queen. As soon as the first of these young queens comes into the world, her natural instinct is to destroy at once every royal cell and its inhabitant, and thus to have no rival to herself. Moved by her jealousy, she tries very hard to do this. And if there are not sufficient bees hatched out to furnish another swarm, or if the weather is unpropitious, the worker-bees allow her to carry out her murderous intention, and, indeed, assist her in the work of destruction. Thus she is left supreme. But, on the other hand, if there are plenty of bees, and the hive is again sufficiently strong to spare another swarm, the young queen, first hatched out, is not allowed to carry out her wish. When she tries to get near a queen-cell the workers prevent and drive her away. At this she becomes excessively angry, and makes a peculiar noise, which may distinctly be heard outside the hive. It sounds like 'Peep,' 'Peep,' uttered harsh and shrill. This is heard by some other young queen yet in her cell, and she also joins in with the same sound, so that it becomes like a challenge to battle given from one to another—the other bees preventing the queen at liberty from taking any unfair advantage over her royal sister in the cell.

When this sound is heard it may be taken as a sure sign of a second swarm in a day or two, for the young queen at liberty, not being allowed to destroy her coming rival, makes resolve herself to leave the hive with as many bees as will accompany her. When this takes place it is said to be a 'Cast.'

Sometimes a third or even a fourth swarm will in succession issue from the same hive.* But these latter

* 'Not unfrequently third, fourth, fifth, and sixth swarms issue from Italian and Syrian colonies, taking the young queens with them. I have taken twenty-seven young queens

should always be prevented if possible, because too weakening to the parent stock. If honey is required, even a second swarm must not be allowed. A great harvest of honey can only be had when the bees do not swarm at all.

With skeps the bee-keeper is to a great extent at the mercy of his bees, and cannot well control these second swarms, but with frame-hives they can always be prevented; for after the first swarm has left, he can take care that no queen-cells remain in the hive, except just that one which is needed to supply a new queen for the hive itself, and without another queen the bees, of course, cannot leave the hive in a second swarm.

Virgil describes another way of preventing a swarm, namely, by clipping the wings of the queen— sometimes practised even now.

> 'The task is easy: but to clip the wings
> Of their high-flying arbitrary kings:
> At their command, the people swarm away:
> Confine the tyrant, and the slaves will stay.'
>
> VIRGIL (by Dryden).

After the first or second swarm has left, if two young queens happen to issue from their cells, as

from a third swarm of Syrian bees. The young queens of these races rarely fight, but live amicably together. I have counted fourteen on a single comb, and the worker bees destroy the supernumeraries after the swarm has issued, sometimes taking a week to complete the slaughter. I am not sure that the workers do not delay the destruction until one of the young queens is ready to become a mother—a further proof of their wonderful instinct.'—REV. G. RAYNOR.

frequently occurs, at the same time, and there are not sufficient bees for another swarm, nothing remains but one of these two rivals must die. They cannot reign together for any length of time. Which is to live, and which is to die? This important question the worker-bees sometimes decide by encasing, and destroying one, and allowing the other to live. But sometimes the queens themselves fight out the matter to the bitter end.

A royal fight has been thus described by Hunter:—'When two queens meet a duel is certain. Like two gladiators, each first takes a good look at her antagonist. Then they rush to the fight. They seize each other by the legs, making with curved abdomen every effort to insert the sting between the rings of the other's body. They wrestle thus, rolling over and over until one succeeds in giving the deadly stroke. It has been stated that if they get in such a position that both are likely to be stung together, they will separate, and commence the fight anew.'

> ' But when two twin-born monarchs burst to day,
> Claiming with equal rights a sovereign's sway,
> Fiercely they rush, unknowing how to yield,
> Where crowds receding clear the listed field.
> Mark how with sharp-edg'd tooth they seize the wing,
> Curl the firm fold, and point the venom'd sting!
> Now, as they view the death-fraught danger nigh,
> With quick recoil, and mutual dread they fly,
> Now, scorn'd all female fears, each hardened foe
> Turns to the fight, and dares the coming blow.'
>
> EVANS.

CHAPTER XXXV.

THE BUSY BEE-KEEPER.

The swarming season over the bees settle down to summer work, and right merrily does it go on when the weather is favourable, when the warm winds blow, and the sun shines, and the flowers are full of honey.

> 'Here their delicious task the fervent bees
> In swarming millions tend ; around, athwart,
> Through the soft air the busy nations fly,
> Cling to the bud, and, with inserted tube,
> Suck its pure essence, its ethereal soul,
> And oft, with bold wing, they soaring dare
> The purple heath, or where the wild thyme grows,
> And yellow load them with the luscious spoil.'
> <div align="right">THOMSON'S <i>Seasons</i>.</div>

And now there is plenty of work for the bee-keeper as well as the bees. Now is his harvest-time, and he must not neglect his busy workers. Now, more than ever, he must do what is wanted without a day's delay. Success very greatly depends upon everything being done at the proper time. The bee-keeper must 'make hay while the sun shines ;' and if he is wise he has prepared things beforehand, and now is ready, as fast as supers are filled, to remove them and to supply others in their place. He will not let his bees remain idle for want of room.

And if he is thus active and careful, and manages well, the quantity of honey he will get, in favourable seasons, is quite astonishing. It is a very common thing to get fifty or sixty pounds of pure good honey in the comb, not only from a single hive, but as an average from all the frame-hives in the apiary. Some experienced bee-keepers take a great deal more than this, even an average of 100 pounds of comb honey, and more of extracted, per hive.*

And all this harvest of honey is of value. If sold it will realise a fair profitable price; but then it must be carefully taken, and neatly put up for sale. Very much of its value depends upon the clean, attractive form in which it is offered, but of this you will learn elsewhere.

But now once more autumn comes, and it is no longer the time of abundant flowers, and the few flowers there are give but little, if any, honey, except in the heather districts. It is the time when people have their holidays, and the bees think they also may have theirs. But they never take holiday, remember, while they have any work to do. As long as there is honey to be had, and room in which to store it, they will work. And thus it is that many bee-keepers find it most profitable, when ordinary country flowers are over, to take their hives to the heaths, if such are near at hand, for the heather gives beautiful honey, and

---

* 'I have had an average of 100 pounds per hive for many years past, and others, I believe, have had as much. I consider that nearly double this quantity can be had by the extractor.'—T. W. COWAN.

flowers much later than other honey-giving plants or trees.

At last, however, even the heather is over, and the bees must rest; and having left nothing to be done at the last moment—as many people do—but having looked forward, and made every preparation, they have both well earned their rest, and can enjoy it.

> 'Oh, Nature kind! oh, labourer wise!
>   That roam'st along the summer's ray,
> Glean'st ev'ry bliss thy life supplies,
>   And meet'st prepared thy wintry day.
>
> 'Go, envied, go, to crowded gates,
>   Bear home thy store in triumph gay;
> The hive thy rich return awaits,
>   To shame each idler of the day.'
>   SMYTH.

The bee-keeper, however, must not quite rest at present, for there is oftentimes much of importance to be done in autumn, especially to those hives which are weak, and short of bees and food, but of all this a guide-book will give information.

There is, however, one operation of which I must here say something, as probably you will frequently hear of it at this autumn season, when it is chiefly practised. It is the operation of 'Driving,' by which the bees in a skep are compelled to come out of their hive, and, leaving all their comb and store behind, to go whither the bee-keeper directs. He knocks at their door. They listen, and with humble submission obey. This sounds marvellous, and indeed appears so to those who for the first time see it done. It calls forth many an exclamatoin of surprise at the bee-keeper's power, and of this it is a good exhibition; but, at the same time, there can be a great deal of

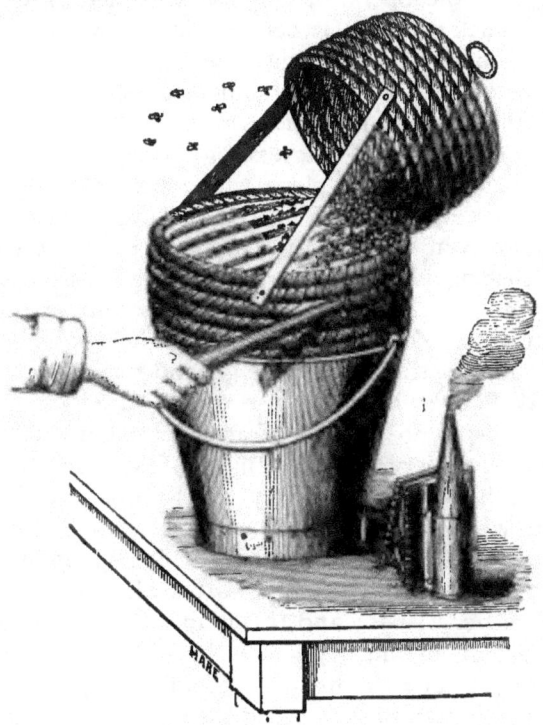

Driving Bees from a Skep.

good bee-keeping without any driving; and with frame-hives it is never practised. The process is as follows:—A few puffs of smoke are given to the bees in a skep. It is then turned upside down, and placed in some position (very often in a pail if the skep has a round top) where it is firmly secured. An empty skep is then fastened to it, as you see in the illustration, and the operator, with both his hands, or two short thick pieces of wood, begins to rap upon the outside of the hive, giving steady blows, sufficient to jar the combs without any danger of breaking them.

This goes on for a few minutes, when the bees become in a state of great agitation, and are seen crowding up to the place where the two hives are joined together. Then in a few moments more they begin to run up into the empty hive, and very soon are rushing up in hundreds with the greatest haste. And now the sharp-sighted operator will see, and, if wanted, be able to capture the queen herself, as, amidst her subjects, she joins in the general rush.

Well, but what is the use of driving? As I have said, it is of no use with frame-hives, but with skeps it is often very useful. In the first place, it is a means by which the honey can be obtained from a skep without killing the bees. And then, amongst other uses, it is the means by which in spring-time skeps can be made to give swarms when and how the bee-keeper wishes; and in autumn it is invaluable in the saving of much precious bee-life, for if the bee-keeper has any neighbours who still cling to the old-fashioned plan of destroying their bees for the sake of the honey, he will persuade them, instead of using the sulphur-pit, to allow him to drive the bees, and to take them home, to join with the bees in his own apiary. He will sometimes even preserve these rescued bees in hives of their own, giving them comb and food, in place of what they had gathered for themselves

During winter there will be little for the bee-keeper to do, and he may 'leave well alone.' But then he must have made due preparation for winter, and, if he has done this, he has taken care to put extra coverings of flannel, with, perhaps, cushions of chaff, on the top of the frames. He has also well packed with chaff or cork-dust the spaces between the two walls of his hives. He will also have taken away some of the frames, and contracted the space in each hive according to the strength of the colony and the number of bees, taking care that the bees fill all the space left them. By all these and such-like means he makes the most of that heat which is necessary to the well-being of the hive.

> 'Thy bee-hives, whether hollowing out of cork
> Thou join'st them, or with rods of osier weavest,
> Construct with narrow orifice ; for cold
> Contracts the honey.'
> VIRGIL (by Kennedy).

The bees have wonderful power to produce and sustain warmth, and to keep the temperature of the hive uniform, but, of course, if there are large vacant spaces containing cold air, and these have to be warmed, as well as the other portions of the interior, there must be great and unnecessary expenditure of heat-producing power, and this means, in other words, a great and unnecessary consumption of food, for with bees, as with ourselves, food is the great means by which the heat of the body is sustained. But as this is a very important subject it will be well to devote a separate chapter to its consideration.

CHAPTER XXXVI.

THE CONNEXION BETWEEN FOOD AND WARMTH.

THE food we eat partly gives us flesh, and partly gives us warmth. It is within us as fuel for what may be called a fire, which, when supplied with fresh air, gives heat to every part of our bodies; and then, if this is so, we can understand that the greater the exertion we make, or the work we do, the more food will be necessary—just as when a train has to go express pace it must have the fire of its engine heaped up with more and more coal. If the fire gets low, and the boiler cold, the train stops. So, when we walk fast and work hard, unless our bodies are properly supplied with the food which replenishes what is lost of both flesh and heat, we become exhausted and waste away, and, if it went on, there would at last be the coldness of death.

Thus the Greenlander, in his very cold climate, and with his very hard work, has need of a vast amount of food, and will eat a quantity of flesh, fat, and oil, which we should think enormous. On the other hand, the natives of the hot climate of India will be satisfied with very much less food—a simple diet of rice, which we should think quite insufficient.

Again, following out the same great law of life, some animals, which are not able to obtain supplies of their proper food in winter, creep into the warmest spot they can find, and there, as it is termed, 'hibernate,'—curl up and go to sleep, and remain perfectly quiet until spring-time comes again. Remaining thus at rest, there is no waste of heat, and all the heat which is needed is obtained from the flesh and fat of the body itself, which becomes during the time more and more exhausted. The bear, which thus hibernates, is at the beginning of the time fat and in good condition, but at the end poor and thin. By keeping perfectly still it has not wasted heat, and it has given up its own fat as fuel for the so-called fire,— its thick fur coat keeping the heat in.

And all this is true of the bee. The bee, however, does not truly hibernate, although, all through winter, it keeps close within, and remains as quiet as possible. Its condition is that of semi-hibernation. And this quietness, coupled with the number of bees crowded together, means plenty of heat. And the greater the number of bees, and the less the space in which they are, means more heat, and this greater heat, thus produced, and thus sustained, causes less food to be consumed.

When there are but few bees, and much empty space with cold air, the poor bees in severe weather, instead of keeping quiet, have to exert themselves, by motion of their wings, in order to give out sufficient heat for the hive, and this, of course, as I have explained, requires the consumption of great quantities of food. Thus when there are few bees, and considerable empty space in the hive, the honey supplies are consumed faster than when there are more bees and less space.

You see thus the reason of what I have described

as the proper management of bees in winter, or rather the preparation to be made for winter—plenty of bees, plenty of food, and warm covering, and no more space than is necessary. Attend to these great principles, and then leave the bees in winter to themselves.

But if it is necessary thus to keep in all the heat possible, perhaps you will ask, ' Shall we close up the entrance, and thus shut out all cold air?' Oh, no! most certainly not. By doing that you would most assuredly kill all the bees; you may contract the size of the entrance, but some amount of fresh air is absolutely necessary. It is indeed by help of fresh air, or the oxygen, as it is termed, in the air, that the animal heat is sustained. The flame of a candle put under a glass case very soon goes out for want of fresh air. Shut off all air from the largest fire, and it will soon cease to burn. And it is the same with ourselves: we cannot live if shut up in a small space without any fresh air. And even in a large room, if it is crowded with people, the windows must be open, or there must be ventilation in some way; and, to a great extent, it is the same with the bees. They must have some ventilation even in winter.

In crowded rooms it is best, because all heated air rises, to open windows high up in the walls rather than those low down; and in some circumstances hives require the same principle carried out. But, generally speaking, the woollen coverings put over the frames will answer every purpose.

CHAPTER XXXVII.

THE BEE-KEEPER IN SPRING.

WINTER well past with the bees, spring comes hopefully. We all feel cheered with its warmer days and its brighter sun, and all nature beginning to burst once more into new life—when 'the flowers appear on the earth, and the time of the singing of the birds is come.' But no one is more cheered in spring-time than the careful bee-keeper. He knows that his bees are ready to take all advantage of spring weather, still strong, and still with plenty of food.

SONG OF THE BEES.

We watch for the light of the morning to break,
    And colour the eastern sky
With its blended hues of saffron and lake :
Then say to each other, 'Awake ! Awake !'
For our winter's honey is all to make,
    And our bread for a long supply.

And off we fly to the hill and dell,
    To the field, to the meadow and bower;
To dip in the lily with snow-white bell,
To search for the balm in the fragrant cell
    Of the mint and rosemary flower.

While each, on the good of her sister bent,
    Is busy, and cares for all,
We hope for an evening of heart's content
In the winter of life, without lament
That summer is gone, or its hours misspent,
    And the harvest is past recall.'

Bee Journal, 1877.

But however well the bees have wintered, there will always be plenty for the bee-keeper to do in spring; only he must not be—as many bee-keepers are—in too great a hurry to do it. Of course with weak hives, and when stores of food are exhausted, he must not delay. Such a state of things—the very sight of empty combs and hungry bees, tells him what to do—that he must at once give them food.

But, with strong hives as well as weak, you must remember that the bees having lived all through the winter will now be comparatively old bees, and will not live much longer; and you must remember what has been said of the great importance of strong hives. And thus we see at once the point of greatest importance to be attended to, namely, that the queen should—as soon as the weather is suitable—begin to lay the eggs, which, producing young bees, shall replace the old ones, whose time for work is nearly over.

And the careful bee-keeper can do something to this end—to hasten the time. It is an important part of his spring work. There is a way of giving liquid syrup which seems to make the bees think that the necessary continuous supply of food for young bees is to be had, and therefore leads the queen to commence egg-laying sooner than she otherwise would. This is called 'stimulative feeding,' but such stimulation should never be practised before there is a good prospect of warm weather. If the bee-keeper is in too great a hurry—many are so—he will do a great deal more harm than good. He must not be beguiled by a few warm sunny days in

February. He must remember that 'one swallow does not make a summer.'

But another thing is to be thought of, because it is not only syrup or even pure honey that by itself is sufficient for the young larvæ. They must have pollen. This is their special flesh-forming food, and, although the bees will have some of it stored up from the previous year, the queen seldom begins to lay many eggs before the workers can find it in tolerable abundance. And so what the bee-keeper does is to give the bees artificial pollen, or something that will answer the purpose of pollen.

This is generally 'pea-flour,' which contains the same flesh-forming substance that pollen does. He places it in shallow boxes near the hives, and it is quite a curious sight to see the bees revel in it, tumbling into it, and getting completely covered with the flour—as white as millers, and carrying it home with the greatest delight.

But nevertheless, although it does well thus as a substitute, the bees, as soon as ever natural pollen is to be had in anything like plenty (very probably from the willows, which flower early), cease to take the artificial food. They greatly prefer nature's supply to anything we can give them. The object, however, has been attained, and the queen has been stimulated to activity, and there will be—earlier than otherwise—plenty of eggs and larvæ in all stages, and many young bees ready to hatch out and strengthen the working power of the hive.

But all this requires care on the part of the bee-keeper, for, as with other things, there is a right and a wrong time to do it. If he does it too soon, and cold weather sets in, the larvæ and young bees will certainly be chilled and die.

And then in spring-time there is another operation, called 'spreading the brood'—gradually enlarging the space of the brood-nest—which is sometimes, when done by experienced bee-keepers, of very great use, and is the means of strengthening the hive with great rapidity; but it is an operation of such difficulty, and requiring such knowledge and care, that I only just mention it. You must not attempt it before you have had long experience.

And now might be mentioned many other things which belong to the spring work of the bee-keeper, but I must leave you to learn of them elsewhere. You will see, however, from what I have said, that it is an interesting and busy season. It is, indeed, quite a time when you must use your head as well as your hands.

A great painter was once asked by a student, who wanted to be saved all trouble in learning his art, 'Pray, sir, with what do you mix your colours, to get these beautiful tints?' To which question there came the gruff answer from the painter, 'Brains, sir. This is what I mix with my colours: brains, sir!' And it was a good answer, full of meaning and good advice. And it is quite one that will do for the bee-keeper. You must help your bees with your brains.

CHAPTER XXXVIII.

DISEASES AND ENEMIES OF BEES.

A WORD now about the diseases and enemies of bees—an important subject, for much will go wrong if we are not careful to watch, and to be ready with remedies and means of protection. And here may be given a very old piece of advice, but none the less useful because it is old,—that 'prevention is better than cure.' And the best 'prevention' possible is, in the first place, not only care and watchfulness, but in an especial manner, cleanliness.

It is the same with the bees as with our own houses and our own persons. To keep free from disease, there must be cleanliness. It is well known that some of the most frequent and fatal diseases, which break out as epidemics, and carry off thousands every year, come entirely from want of this care and cleanliness, especially the want of pure water and good drainage. Many diseases will, of course, come, notwithstanding all care, but very often they are thus preventible diseases, and may be kept at a distance by the exercise of forethought and care.

So with the bees. Due care and attention to cleanliness, and watchfulness against all causes of disease, and being ready, when disease first begins, to ' nip it in the bud,' will do very much to keep our bees healthy.

I cannot now name all the diseases, but I may say

that the most fatal of all is one called 'foul brood,'—when the young brood die and rot in the cells, and which is not only a fatal but also a most infectious disease.

It has often ruined whole apiaries, causing most serious loss. Many remedies have from time to time been tried, but only with partial success. Now, at last, what appears to be a sure remedy, an old remedy in a new form, has been discovered, so that, we hope, it will no longer be the dreaded pest it has been.

And what I have said of care and watchfulness in the matter of disease, will equally apply as useful advice with regard to many of the enemies of bees.

One of the most serious of these enemies is the wax-moth, which is particularly fond of laying its eggs in any crevice in the hive, and the grubs from which are most obnoxious and destructive. But these eggs may be looked for and destroyed—at least with frame-hives, although with skeps this is impossible.

> 'Their chambers oft
> Are choked with skulking beetles. . . . . . .
> Or moths, an execrable race, intrude,
> Or savage hornet, with unequal arms,
> Or spider, hateful to Minerva, hangs
> Her straggling network at the vestibule.'
> VIRGIL.

Other enemies are mice, slugs, and snails, but very ordinary care will prevent much danger from these.

Some birds are also enemies. The blue tomtit,—pretty little bird as it is,—is especially so in winter and early spring; for, when once it has had a taste of bee-flesh, it will again and again come to the hive for

DISEASES AND ENEMIES.

fresh supplies, and, tapping at the door to draw the bees out, will seize them as they make their appearance, and carry them off to some neighbouring tree, there to eat them at leisure.

> 'These rob the trading citizens, and bear
> The trembling captives through the liquid air.'
> VIRGIL.

Far worse enemies, however, are the wasps, for when once they have obtained a taste of the good honey within, they will, with great perseverance, force their way into the hive, and being active, strong, and resolute, will cause a great deal of mischief.

But here, again, much can be done in the way of prevention by carefully destroying queen wasps in the spring, and wasp-nests later on; and also by lessening the entrance to the hive attacked, and so giving the bees more opportunity to defend it. Wasps, however, but seldom attack a strong hive, and thus here, again, we see the importance of the golden rule, 'Keep your hives strong.'

But even yet worse than wasps are robber bees. The bees of any neighbouring hive, when once they begin a thieving life, are the most desperate thieves known.

> 'They muster all in haste, their pinions flash,
> Their stings they sharpen, and adjust their claws.'
> VIRGIL.

They will without pity attack a weak hive, and when once they begin their depredations are most difficult to subdue or stop. Virgil says—

> 'All this commotion, all this deadly fray,
> The scatterring of a little dust shall quell.'

But I am afraid this will not always be quite effective. And what is worse, this ill conduct, as usually the case with bad examples, is most infectious, so that very often the whole apiary joins in the raid, and is, for a time, thoroughly demoralised. But here, again, carelessness is often the cause. A little honey, left exposed near the hives, will often raise the storm.

CHAPTER XXXIX.

THE USES OF HONEY AND WAX.

I MUST now tell you something of the great usefulness of bees. Possibly you think that their only use is to collect honey and to make wax. But this, indeed, is very far from the case; it is not one half the truth. We will try and see how this is; but it is a very large subject, and I can only give you the most general outline, but sufficient, I hope, to make you wish to know more, and to see more clearly how marvellous and wise are all the ways of that Providence which rules in nature.

But, first of all, I must say a word of the honey itself. This, of course, is useful, and we keep bees, in a great measure, for its sake. You like it; it is pleasant to the taste. In olden days, and, indeed, until something like four hundred years ago, it was invaluable as the great substitute for sugar. Virgil speaks of its use :—

> 'T' allay the strength and hardness of the wine.'

And then, in those days gone by, several beverages, and some fermented drinks, were largely made of it. Mead is often made with it now. In Russia a drink is made of it of about the same strength as beer, and is an article of regular consumption. It is called 'Mjod.'

But it is as actual food that it is also very useful; I mean by this that it can not only be used largely in many articles of food, but that it has in itself some of those properties which supply the wants of our bodies.

There is with our bodies a daily wear and tear going on, which soon causes exhaustion, unless they are continually replenished with proper food. And the food we take for this purpose must be of varied kind; for one description of food, or some portion of that food, supplies a certain class of wants; and another food, or a part of such food, some of the other wants. No one food by itself, except, perhaps, milk, supplies all the wants of our bodies. Roast beef is good, but no one could live upon it without something else. The body would soon starve, for, although the beef contains some things good for the body, it is wanting in others which are essential. Thus, as I have said, there must be variety in our food.

One of the most important varieties which is necessary to us is sugar. I do not mean only the sugar as you generally eat it, but the sugar which, although you do not know it, is contained in many things which you eat. A great deal of this important sugar is contained in honey, and is there present in a form which makes it especially useful to the body, giving it heat and energy, and acting as a most useful stimulant. It supplies, in short, some fuel to that fire of which I spoke in a previous chapter. We could, of course, no more live upon honey than we could upon beef, but, as a variety of food, it is thus most useful; and may be made to enter most advantageously, as well as agreeably, into the manufacture of a great number of articles of food.

Then, again, speaking of the usefulness of bees, they are very useful because of the wax which they make. Wax is a most important product, and is put to many uses.

Formerly many hundreds of tons of it were used to make candles, and some is even now thus used. In Spain bees are kept in some parts almost entirely to provide wax for tapers used in the Roman Catholic churches. In Russia, also, only such candles as are made of pure bees-wax are used in the churches. But since so many kinds of cheap oil have been discovered, and gas has come so extensively into use, bees-wax has been in most countries, in great measure, superseded.

It is still, however, much used for other purposes, such as polishing and cleaning, and as an ingredient in some articles of manufacture. A vast quantity is also now used in making that comb-foundation which is so invaluable in modern bee-keeping. The wax is thus given back to the bees, and is used over and over again for the purpose of fresh comb.

And not only honey and wax, but even propolis has some use as a product of the bees. It is extensively used in Russia for lacquer-work. It is also said to make an excellent glue for some fabrics for which ordinary glue does not answer.

And here, while dwelling on the uses of bees and their products, I may tell an amusing story of a very unusual, but ingenious, use to which some bees were put on one occasion many years ago in Austria.

For some reason there was a great uproar in a certain town, and a very angry mob collected, surrounding the house of the chief inhabitant. They threatened violence. They would not listen to reason. They were about to attack the house. No time was to be lost, but no help was at hand. What could be done? Both house and owner were in danger. But a happy thought, and just in time, occurred to the latter. He called his servants, and told them to run and bring him his hives of bees as quick as possible. And these at once, with all his force, he threw amongst the crowd. You may imagine how the bees rushed out, and began to attack everyone near. At all events, it fully answered the purpose, and far better, and quicker, than any good advice. It was more than the crowd could stand, and in a few minutes they fled, scattering here and there, to escape, if possible, the stings of the countless and enraged insects.

But leaving all such uses of bees, we must pass now to the consideration of that part of the subject of which I wish to speak more particularly, and which, when understood, shows that the bees are not only the most wonderful, but the most useful, of insects.

CHAPTER XL.

FLOWERS IN RELATION TO BEES AND OTHER INSECTS.

OF course you know that bees could not exist without flowers; but do you know that many flowers could not exist without bees?

Now it is this that I want to show you. I want you to have some idea of the great truth, that bees are quite as useful to the flowers as the flowers are to the bees, and that, if it were not for the bees, a great many flowers would altogether die away, and a great deal of what is beautiful around us would be quite changed in appearance. And further than this, many of our valuable fruit-bearing trees would, without bees, almost cease to give us any crop at all. They might flower, but they would not bear fruit. This seems a great deal to say, but it is quite true. In order in any way to understand it, you must, however, first know a little of the construction of a flower, and what a flower requires.

What, then, is a flower made of? What are its parts? There are, of course, numberless forms of construction. Nothing varies more than the form, colour, and appearance of flowers; but if we take them to pieces, we shall find in all, to some extent, the same parts, each designed for its own proper purpose.

Suppose, for instance, we take the flower of a cherry-tree. Pick it carefully to pieces, and you will find that you can separate it into many parts, all of which have names. I will not, however, now trouble you with all these names, but some of them you must remember in order to understand what I am going to say about bees and flowers.

Well, first of all, take away the white outer leaves of the flower, which are called *petals*. When they are removed, you have in your hand a small, solid-looking centre, from the middle of which springs up a thread-like little stalk, which, remember, is called the *pistil*.

Take notice of this pistil, and examine into its origin, and you will find that it leads down to a receptacle called the *seed-vessel*, in which is situated the very tiny thing which, when the time comes, will grow into a seed, and ripen ready for another year. At the upper end of the pistil there is a sticky substance called the *stigma*.

But now look again at the flower, and around the pistil you will see several other fine, thin stalks called *stamens*, having at their upper ends not a sticky substance like the stigma, but a very small, oval-shaped head, which, when the time comes, and it is ripe, will develope into a small packet of the finest possible dust, which is the *pollen* of which you have heard so much.

This is the simplest arrangement. You will understand it by looking at the first of the following illustrations, which is that of a cherry blossom cut in half. You will see the same in rather a different form in the apple blossom, and the same in another simple form in the buttercup.

And now, whatever the plant, or the construction

of its flower, you must always remember this most important fact—the great law of life in all flowering plants—that before that tiny little beginning of a seed, of which I spoke, can come to perfection, and be a ripe seed, *some small portion of the fine pollen-dust from the end of a stamen must fall, or be placed, upon the stigma of the pistil.* Unless this happens, the flower fades and falls, and no fruit or seed follows. When it does take place, then the great object of the flower is attained. The flower is 'fertilised,' as it is termed, and, in the case of the cherry, just for one instance, the flower gives place, after a time, to the luscious fruit, which encases the stone, which, in its turn, is the hard, protecting case of the kernel or true seed. In the first condition of the flower this kernel, or true seed, was the tiny thing of which I spoke, situated at the lower end of the pistil.

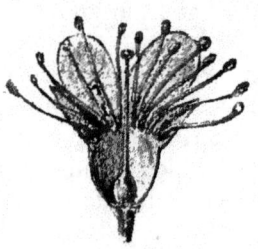

Section of Cherry Blosoms.

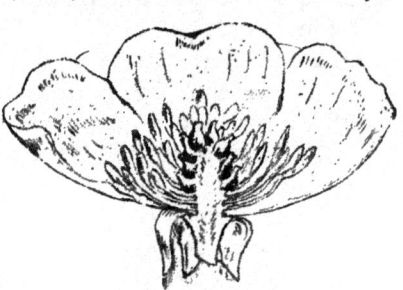

Section of Buttercup.

We must now, however, proceed a step further, and take notice of this, that before any such result can come from the pollen being placed upon the pistil, it is necessary that both the pollen and the pistil should be ripe at the same time. There is, in fact, just one right time, and no other, when that which is necessary can take place.

Section of Apple Blossom.

If it happens in a flower that both the pollen-bearing stamens and the stigma-bearing pistil are ripe at the same time, and there is no peculiarity of construction in the flower to prevent it (although there very often is), the process is easy. The wind, perhaps, shakes the flower, and a little of the pollen-dust is blown on to the pistil, and, since the stigma is sticky, adheres to it, and nothing more is needed. Nature does the rest.

But with a great many flowers the process is not so easy. Nature, indeed, for a most wise purpose, puts various difficulties in the way. For instance, the fine pollen dust of some flowers is so constituted as to be of no use to the pistil of its own flower,—although both may be ripe together, and may come in contact, —but can only fulfil its purpose when carried to another flower, or, in some cases, to another plant of the same species. It is so remarkably with the common primrose. Again, in a great number of flowers, the stamens are not ripe at the same time as the pistil. In some the stamens are ripe first, and shed all their pollen, and then afterwards the pistil appears, and holds up its head ready for pollen, which now its own flower cannot give it. In other cases the pistil is ripe first, but cannot get pollen from its own flower, as the stamens are not as yet in a sufficiently forward state.

Here is a meadow geranium. When, first of all, the stamens are ripe the pistil is not ready. You see this condition in Fig. 1. After all the pollen has been shed the stamens die away, and the pistil appears as you see it in Fig. 2.

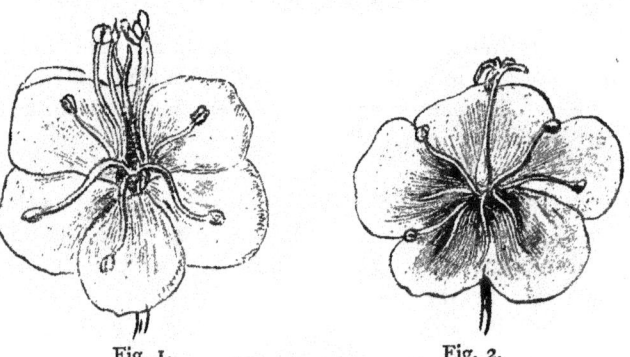

Fig. 1.   Fig. 2.
Meadow Geranium.

As another example, we have the wood sage. In Fig. 1 you see the stamens standing forward, and the pistil behind, not yet ready for pollen. Afterwards, when the pollen has been shed, the stamens bend down, of no more use, and the pistil stands erect, ready for the pollen from some other flower, as you see in Fig. 2. When the bee visits the first flower, it does not touch the pistil, but only the stamens and pollen. In the second stage it only touches the ripe pistil.

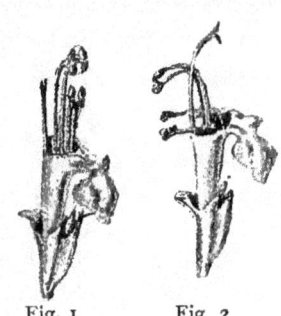

Fig. 1.   Fig. 2.
Wood Sage.

Again, in other flowers, although stamens and pistil are mature together, the pistil is so situated, at the time, that it cannot get the pollen belonging to its own flower.

Again, some plants have two kinds of flowers, the one kind with only the stamens, and the other with only the pistil.

And, once more, of some kinds of flowers, one whole plant will have only stamen-bearing flowers, and another none but pistil-bearing flowers.

So that, as I said, there seems to be all kinds of difficulties,—and there are many more than I have mentioned,—in the way of getting pollen to its appointed place; and yet it must get there.

What is the way out of the difficulty? Nature has made the difficulty, how does nature provide for it to be overcome?

Well, first of all, you must remember that, although in any one flower the stamens and pistil may not be mature at the same time, yet that there will certainly be close at hand, perhaps on the same plant, other flowers in which there will be the mature pollen or the mature pistil, as the case may be, or, in fact, just the state of things required, if only the pollen can be carried and placed where it is wanted. If, in any flower we may be looking at, we see that the stamens have died away, and the pistil is ready and ripe, we may be almost sure that, somewhere near, there is a flower which is not in such a forward state, but where only the pollen is ready.

And now, I think the great truth, to which all this leads will already have occurred to you, and when the question is asked, 'How can the mature pistil get the mature pollen which it requires?' you are ready with the answer that it is done, not only by the wind, but by the work of bees and other insects, which when they come searching for honey, get the pollen on to their bodies, and so carry it from flower to flower.

Yes, this is the great truth; and the honey,—which is situated in what is called the 'nectary' of the flower,—is by the wonderful care of Nature so situated (its situation being varied according to the form and situation of the stamens and pistil), that, when a bee gets into the necessary position on, or in the flower, in order to gather the honey, its hairy body must of necessity touch, in the first place, the ripe pollen on the stamens, which at once, like dust, sticks to its hairs, and so is carried away, and then rubbed off again by the ripe pistil of another flower, which,—on account of the position of the nectary, cunningly situated,— cannot fail to touch the body of the bee as it makes an effort to obtain the honey.

The bee thus does the very thing which is needed to cause the seed of the flower to come to perfec-

FLOWERS IN RELATION TO BEES.

tion. It is Nature's handmaid and most invaluable servant.

Just for one example of the marvellous way in which provision is made for this law of plant-life to be carried out, I may point you to the flower of the

Bees and Orchis.

orchis, where the pistil is so situated that the only way in which the pollen can get to it is by its being carried on the bee's head.

The bee goes to the flower, and, while it is busy

getting honey, it displaces by its head—the nectary being situated so that this takes place—a little thing which looks like a couple of horns coming out of a little sticky crown, and so displaces it that the crown actually sticks to its head, as you see in the drawing on preceding page.

Crowned with this the bee flies away, looking like a bee with horns, but by the time it has got to another flower these little horns have so altered their position that, when the bee inserts its head into a fresh flower, these little horns,—which are really stamens with pollen attached,—go exactly to the very spot where the pistil is waiting for them; and thus the pollen is placed upon it, which could not be the case in any other way.

This, however, is only one example, and the whole subject is full of such marvels; but sufficient has been said for our present purpose, sufficient to show the usefulness of the bee, for, although the wind does much, and a vast number of other insects—moths, butterflies, and many more—join in the same work, no one kind does so much as our friends the bees.

But perhaps you will say, 'I wonder why it is that all this trouble is necessary. I wonder why it is that every flower is not complete in itself, and has not stamens and pistil ripe at the same time.' And this of course would seem the simplest plan; but, although I cannot explain it all, I may say that the arrangement, as it is, is one of the most wonderful, and one of the most striking displays possible of that wisdom which is seen in all Nature.

When, under certain circumstances, flowers fertilise

themselves, their colour and beauty become less and less marked, the plants themselves degenerate, become poor and weak, and the whole race is in danger of extinction. And thus, even in those cases where the whole of the flower is ripe together, it is of the greatest importance that visits of bees and other insects should take place, for the pollen is thus carried about from flower to flower, and no one flower lives and dies, as it were, by itself. Every flower in some way helps and benefits its neighbour.

And now, further, from what has been said, we can see more clearly the important use, not only of bees, but of honey. It is, we know, very useful to ourselves, but we can understand now that it is much more useful to the flowers,—their very existence depending upon it, for, were it not for the honey, bees and other insects would not visit the flower, and it would remain unfertilised, and thus would never have any ripe seed. The flower holds out the attraction of its sweets; it invites the bee : 'Come, and I will give you my honey, and you, in return, shall bring me the pollen I want; and then shall take of mine, and carry it to other flowers now in need of it.' And thus the great purpose of Nature is fulfilled.

The colour of the flower also does its part in the work, according to the law that everything in Nature has some object and reason of existence. You remember that I explained how we know that bees have some knowledge of colour, and so are attracted to the flowers by the colour, as well as by the honey. And, although many flowers, without any beauty of colour, give abundance of honey, it is a remarkable fact,—

whatever the explanation of it,—that in the case of all those trees and plants which do not require the visit of any insect, but only require the wind to scatter the pollen, the flowers are without colour, without scent, and without honey. They do not need the insect, and so do not hold out any such attraction or give any such invitation.

Some day you will be able to understand more of this marvellous and interesting subject, and will learn more and more the lesson of Divine truth which it teaches, but, even now, reviewing the mere outline of the subject which I have given, you cannot fail to be astonished at the great work done by the bees, and to see their vast usefulness, usefulness so great that it has been well said by one of the greatest authorities on the subject,* 'To them we owe the beauty of our gardens and the sweetness of our fields.'

> 'There are in these examples who discern
> Proof that in bees a power ethereal dwells,
> An inspiration of the soul divine.'
>
> VIRGIL (by Kennedy).

CHAPTER XLI.

THE IMPORTANCE OF BEE-KEEPING.

BEARING in mind what has been said in the last and other chapters of the great usefulness of bees,—the honey they collect, the wax they make, and the flowers they fertilise,—we can understand that it is of considerable importance that as many as possible should be kept everywhere.

* Sir John Lubbock.

IMPORTANCE OF BEE-KEEPING.

It has been said by some that if a great number of stocks are kept in any one place, the flowers in the neighbourhood will not furnish sufficient honey. But this is quite a mistake, except perhaps as regards a few very barren localities. Some of the greatest authorities assert that no district has ever yet been overstocked. Langstroth says, 'It is difficult to repress a smile when the owner of a few hives, in a district where as many hundreds might be made to prosper, gravely imputes his ill success to the fact that too many bees are kept in his vicinity. If in the spring a colony of bees is prosperous and healthy, it will gather abundant stores in a favourable season, even if hundreds, equally strong, are in its immediate vicinity, while if it is feeble it will be of little or no value, even if it is in "a land flowing with milk and honey," and there is not another stock within a dozen miles of it.'

In fact, almost any number of hives may be kept, and will give good return if managed well. At all events, in other countries, many more are kept than in England. In some places there are apiaries containing several hundred stocks, and even as many as two or three thousand.

In Russia, Germany, Austria, Greece, Cyprus, and many other countries, bee-keeping is practised most extensively—far more so than in England—and for a long time has been a national industry—a position which it has only very recently assumed with us. Langstroth mentions that 'in the province of Attica, in Greece, containing forty-five square miles and 20,000 inhabitants, 20,000 hives are kept, each yielding on an average thirty pounds of honey, and two pounds of wax.' And, again, that 'in 1857 the yield of honey and wax in the empire of Austria was estimated to be worth over seven millions of dollars.' And that 'a province of Holland, containing 1200 square miles, maintains an average of 2000 colonies per square mile.'

But it is in America that bee-keeping is now carried on most extensively. In that great country, where, owing to the difference of climate of the vast tracts of land through which some great river passes, the flowers of several districts bloom in succession, the plan has been tried, although not with much success, of placing many hives on a barge, which night by night is towed to fresh pastures, where the bees roam by day, and then, returning to the hives at evening, are carried on to fresh fields before the next morning. We read of even 400 or 500 hives, being so placed on a couple of barges, and towed by a steamer up the river from New Orleans. Not that this is altogether a new plan, for it was practised to some extent even in ancient Egypt; and in France and elsewhere has been an old custom, showing, if not an example for ourselves in this country, yet, at all events, a proof of the enterprise of foreign bee-keepers.

In America, again, bee-farms are established on the largest scale, and are managed on the most scientific principles. A large apiary in California is said to have given sixty-seven tons of surplus honey in one year, and an apiary of 500 stocks is by no means an unusual thing.

And if such things are possible elsewhere, and there is honey sufficient for such large numbers, we may safely say that oftentimes there are hundreds of tons of honey in our fields and orchards that are wasted. The apple and other fruit-trees in the spring, the bean and clover fields in the summer, and the lime-trees later on, to say nothing of hundreds of other kinds of flowers, are all full to overflowing of precious nectar, but there are not bees to gather it. And it is not as if these flowers could give up their supply only once, for nature is so bountiful that, as mentioned before, bees may come again and again, and find the little storehouse replenished. It is the voice of the flowers to the bee, 'Take away all my sweets, and yet come again, for some more will very soon be ready for you.'

What a pity it seems that there should be such waste of Nature's good gifts! It is waste also of that which might bring many a comfort to those who need. The cottager, who oftentimes has hard enough work to make ends meet, and can only do so by great care, might very well add something to his store by keeping a few hives. It would cost him but little time, or only such time as his wife or elder children could give. We may almost say that the money for his rent, or the money for his children's shoes, is lying there in the fields, every flower containing its little mite ready for him, if only he would keep the busy workers who are ready most willingly to gather it for him.

And, besides all this, it is doubtless true that, if more bees were kept, there would be yet greater crops of good fruit in many an orchard and garden. We see an orchard white with lovely blossoms, but a vast number of these fall to the ground, and never develope into fruit for want of the visit of a bee. The wind does much, and many other insects help, but of all workers amongst the flowers, the bee, we know, is first and foremost and most useful. And so much is this the case that when fruit-farming is undertaken and large orchards are planted, some people very wisely establish at the same time large apiaries, not so much for the honey the bees produce, as for the good they do amongst the fruit trees. On every ground then we bid 'welcome to the bee.'

WELCOME TO THE BEE.

'Come, honey bee, with thy busy hum,
To the fragrant tufts of the wild thyme come,
And sip the sweet dew from the cowslip's head,
From the lily's bell and the violet's bed,
  Come, honey bee,
  There is spread for thee
A rich repast in wood and field :
  And a thousand flowers,
  Within our bowers,
To thee their sweetest essence yield.

Come, honey bee, to our woodlands come ;
There's a lesson for us in thy busy hum ;
Thou hast treasure in store in the hawthorn's wreath,
In the golden broom, and the purple heath ;
  And flowers less fair
  That scent the air
Like pleasant friends, drop balm for thee,
  And thou winnest spoil
  By thy daily toil
Thou patient and thrifty and diligent bee.

We may learn from the bee the wise man's lore,
"The hand of the diligent gathereth store :"
She plies in her calling from morn to night,

Nor tires in her labour nor flags in her flight:
From numberless blossoms of every hue,
She gathers the nectar and sips the dew.
    Then homeward she speeds
    O'er the fragrant meads,
And she hums, as she goes, her thankful lay:
    Let our thanks, too, rise,
    For our daily supplies,
As homeward and heavenward we haste on our way.'
Bee Journal, 1877.

CHAPTER XLII.

SUPERSTITIONS WITH REGARD TO BEES.

THROUGHOUT this book I have endeavoured to point out from time to time that for everything in the natural history of bees, their structure and their habits, there is good and sufficient cause. We thought, for instance, of the tongue and the sting, and of their legs and wings, of the young they rear, and the combs they build, and each bee doing its own appointed work; and we saw wise purpose in everything.

In this chapter I will speak not of the bee's wisdom but of man's foolishness. I will tell you of one or two curious things, sometimes said to be true, by those who are ignorant and superstitious, but for which there is but little foundation of either truth or reason.

'Bad luck,' as it is called, in bee-keeping, is connected by superstition with many things which can by no possibility have anything to do with failure. Very likely, in years gone by, people, as they do now, lost their bees through bad management, and then, remembering something that had happened at the same time, began to think, without any reason whatever, that the two things were connected together. And when perhaps, by a curious coincidence, the same two things happened again at the same time, the idea began to take root, and so grew, as superstitions will grow, until at last people became quite persuaded that if such or such a thing happened it would certainly bring misfortune to the bees.

With some foolish or ignorant people it is considered very unlucky to walk under a ladder. Of course it may be dangerous, but 'unlucky' has no meaning. With others there is a superstitious belief that if thirteen persons sit down to dinner together one will die before the year is out. Many foolishly think it is unlucky to spill salt at table; others that it is unlucky and portends a death if a dog howls at night near a house. Sailors think it unlucky to set sail on a Friday.

And then with bees. There is a common superstition that it is most unlucky to sell them except for *gold*. Hence in many parts the usual price of a swarm is half-a-sovereign, to be paid in gold. Again, some people think it unlucky to begin bee-keeping by buying the bees of their first stock. If they are bought they will do no good; the stock must be a present, or a swarm must come of its own accord. I remember that when I began bee-keeping many years ago I was told that I should be lucky because my first stock was a present from a friend.

Another curious idea prevalent in some parts of the country, as told by Harris, is that if **bees**, when they swarm, alight on dry or dead wood—a dead bough or a post, they will never prosper. He also quotes an old author, who says, 'The Cornish to this day invoke the spirit " Browny " when their bees swarm, and think their crying " Browny! Browny!" will prevent their returning to their former hive, and make them pitch, and form a new colony.'

Closely allied with these superstitions as to swarming, is the old and very prevalent custom of making a clatter called 'tanging,'—if possible with a key upon a tin pan,—while the swarm is in the air, for the purpose, it is said, of causing it to alight near at hand, and not to fly away.

This 'tanging' is a very ancient custom. Even Virgil,—and he takes the idea from earlier writers, —mentions it as very advisable. Sweet-smelling savours are to be scattered on the ground, and then,

> 'Make tinkling noise and beat the Phrygian drum;
> They of themselves, attracted by the scent,
> Will settle, and in fashion of their own
> Take full possession of their infant realm.'
>
> VIRGIL (by Kennedy).

And it has probably been more or less a custom ever since, and even now is practised in many parts of our country, although for its professed object it is perfectly useless.

In this case, however, the old custom may have survived owing to the noise being taken as loud notice, given by the owner of the swarm to all his neighbours that his bees had swarmed, and thus as a claim of his right to follow them and to secure them wherever they might settle. It is said that 'by one of the laws of Alfred the Great all bee-keepers were bound to ring a bell when their bees were swarming, to give notice to their neighbours of the fact.'*

But by far the most curious superstition is that which supposes some mysterious sympathy between the bees and their owner when any death occurs in the family, and which makes it necessary to inform the bees of the event, and to make them share in the mourning. A writer in the *Bee Journal* says, 'A person told me a few days ago that her grandfather, who lived in Oxfordshire, had seventeen bee-hives, and when he died the bees were not informed of it, and the consequence was every one of the bees died.'

The usual practice is to 'tell' the bees of the death at midnight by tapping the hives and saying So-and-so is dead,' and afterwards to pin a piece of crape on each hive.

And this old superstitious custom still lingers in all parts of the country. Only the other day (1885) I saw hives in Norfolk thus in mourning; but it is of course a custom that will gradually die out as people more and more learn the true principles of bee-keeping, and the true reasons of success or failure. The Rev. George Raynor, the well-known bee-keeper, has given me the following story:—'An old "lady" in this parish, whose husband died a short time ago, was about "to put her bees into mourning," when I dissuaded her, showing her how foolish was the idea that the bees could understand anything about the death. During the following winter the

* *Bee Journal*, vol. iv.

bees died. I was never forgiven, although I offered more bees to the manes of the departed husband. I was greeted with, "It's all very good o' you, sir, but they ain't like t'other poor dears as is dead and gone!"'

CHAPTER XLIII.

BEE-KEEPERS' ASSOCIATIONS AND SHOWS.

FOLLOWING up what I said in the last chapter, we may consider it as a general truth, that what is called want of luck is really want of management, and most certainly can have no connexion with any of the several causes assigned by superstition. There is an old saying that 'a bad workman complains of his tools,' and it is often so with the bee-keeper. He blames anything rather than himself. Bad weather may, of course, bring unavoidable misfortune, but even this can in great measure be averted by good management.

And now in these days there is less excuse than ever for any one to talk of ill-luck, for if a person really wishes to succeed, and is willing to learn, and ready to take trouble, there are abundant means at his disposal for learning all that is necessary. There are not only books and guides and periodicals in abundance, but, most probably, there will be a bee-keepers' association in his county or district; and, if he joins it, he will by his membership obtain many benefits and much information.

These associations established in almost every county in affiliation with the British Bee-keepers' Association, do a great deal to advance intelligent bee-keeping. They not only promote unity of purpose and a kindly feeling amongst bee-keepers, but in various ways give much instruction. They organize shows, and give prizes for honey and hives. They provide lecturers for meetings, and then, as their principal and most useful work, employ men of experience called 'experts,' who go round the country, and visit members at their homes, examining their hives and giving good practical advice.

When you are a bee-keeper you will doubtless join such an association, and then, when your bees are prospering, you will look forward to its annual show with great interest. It may be you will be able to send some honey for exhibition. Let us hope you will.

The show of hives and honey will possibly be in connexion with the flower show of the neighbourhood, so that there will be a large gathering of people from the county round—those interested in flowers as well as those interested in bees.

The occasion of such a show is both an interesting and pleasant holiday. It will be held, probably, near the market town, and you will perhaps go by train; and as you draw near, at every station you will see signs of holiday and preparation. Children with their exhibits of wild or other flowers will join the train, dressed in their best, and with anticipation of success beaming in their faces. Gardeners, too, will be there with their precious flowers, eagerly discussing their merits.

And then others will enter the train, bringing their exhibits of honey—the supers their bees have filled, and which for long have been the objects of much anxiety. Carefully and neatly packed they will be eyed once more, and again, by the exhibitor, and the story will be told to fellow-travellers of the wonders the bees have performed, and how the honey they have made can hardly fail to win a prize.

And thus at last the show-ground is reached—a spacious field or the shaded recesses of a gentleman's park, brightened with flags and banners. Here all is bustle and preparation amidst the several tents,—one for cut flowers and hardy plants; one for ferns and hothouse plants; one, perhaps, for roses, the queen of flowers; one for cottagers' exhibits, and then, last, but not least, one for the exhibits of hives and honey. Carts and vans have brought heavy loads, which are now being transferred to the appointed tents. And here you give up to the Secretary or Manager your own valued contribution,—well-filled sections, or large supers, or bottles of clear run honey, to be arranged in due course, each in its own class, there to be inspected by the appointed judges.

And now you must wait; and anxious the time of this waiting, while the judges make their awards. The time, however, comes at last, and the tent is thrown open. Exhibitors and visitors press in. And great your pleasure, as you enter with the throng, if you find 'First Prize' or 'Second Prize' on your exhibit! It is reward for your care. But if no prize is yours—and all cannot win—let us hope you will bear the disappointment bravely, ready to approve the judges' decision, and making the determination that you will not be beaten another year, but will yet succeed by perseverance and greater care.

> ''Tis a lesson you should heed,
>   Try, try, try again.
> If at first you don't succeed,
>   Try, try, try again.
> Then your courage should appear;
> For if you will persevere,
> You will conquer, never fear:
>   Try, try, try again.
>
> Once or twice though you may fail,
>   Try, try, try again.
> If at last you would prevail,
>   Try, try, try again.
> If we strive 'tis no disgrace
> Though we may not win the race;
> What should we do in that case?
>   Try, try, try again.
>
> If you find your task is hard,
>   Try, try, try again.
> Time will bring you your reward:
>   Try, try, try again.
> All that other people do:
> Why, with patience, should not you?
> Only keep this rule in view—
>   Try, try, try again.'

And now, as you walk around the show, keep your eyes and ears open to gain all possible information. You will meet with many of greater experience than yourself—the best bee-keepers of the county, and you will always find them ready to give you kind words and advice. Bee-keepers, like their bees, try to help one another, and to work for one common end.

A Bee Tent.

And then, having thoroughly examined all the exhibition tent contains, you will visit the 'bee-tent,' where, safe from attack behind the gauze net, many will be gathered to witness some expert engaged in driving bees, and transferring them from a skep. You will hear him also explain some of the wonders of the hive, and the best way of practical management. Observe and listen carefully, and you will pick up many a hint to be brought into practice afterwards.

And thus, amidst beautiful flowers making the air sweet, and amidst bees, and honey, and pleasant friends, and cheerful faces, and in listening to the band, you will spend a happy day, and, when evening comes, will return once more to your home and hives, all the better for having mixed with others, and gathered experience, and all the more, I think, loving your bees because understanding them better.

And this interest will not flag in future years. On the contrary, I think, as years pass away, you will be all the more attached and devoted to bee-keeping. There will be ever a growing interest, not simply for profit sake, but because you will ever find more and more to learn, fresh wonders from time to time, wonders of bee life, wonders of bee instinct, wonders amidst the honey-giving flowers, each one more clearly unfolding the wisdom and goodness of the great Creator, who desires that we should enjoy to the full the rich and countless gifts of Nature so freely bestowed upon us all.

LONDON:
Printed by STRANGEWAYS & SONS, Tower Street, Upper St. Martin's Lane.

2 Paternoster Buildings, E.C.

Wells Gardner, Darton, & Co.'s
Catalogue of Books.

ADAMS.—WORTHIES OF THE CHURCH OF ENGLAND: A Series of Biographies of Priests and Laymen of the Church of England. By W. Davenport Adams. Crown 8vo. extra, cloth boards, 3s. 6d.

AINSLIE.—INSTRUCTION FOR JUNIOR CLASSES IN SUNDAY SCHOOLS. By the Rev. A. C. Ainslie, M.A., Vicar of Langport, Somerset, Prebendary of Wells. Fcap. 8vo. cloth boards, 1s. each.

Vol. I. The Story of the Gospels in Fifty-two Lessons.
Vol. II. Fifty-two Lessons on the Acts of the Apostles.

These books are specially designed to help those who have not been trained as Teachers. For the convenience of Teachers, a packet of Lesson Leaflets for a class of ten, price 5s.

ARMSTRONG.—THE KING IN HIS BEAUTY, and other Poems. By Florence C. Armstrong. With Outline Illustrations by H. J. A. Miles. Square 16mo. cloth, bevelled boards, gilt edges, 1s. 6d. [Second Edition.

ANDERSEN, HANS.—THE SNOW QUEEN. See Pym.

THE ARTIST: A JOURNAL OF HOME CULTURE.
For Artists, a complete monthly professional newspaper.—For Architects and Decorators, a Monthly Magazine for the artistic side of Architecture and House-fitting, under the sectional title of 'The Architect and Decorator.'—For Dealers and Collectors, a regular chronicle of Art Sales, showing the state of the market for Pictures, Porcelain, Art Books, Engravings, Old Furniture, and Bric-à-Brac.—For the Potter and China

THE ARTIST.—(*Continued.*)

Painter, a section entitled 'Keramics.'—For Engravers and Etchers, a department entitled the 'Engraver and Etcher,' with notices of New Engravings and Etchings. — For the Art and Fancy Trades, a section for noting the development of Art in Furniture, Stationery, Household Articles, Art Apparatus, &c.

'*The Artist*' provides notes on novelties of taste for the house, the table, and the garden, and on Art Work for Ladies. Other features are Local Art Notes, a review of the month's Music and Drama, a department for Photography, papers on Art Abroad, Foreign Correspondence, Art Literature, Queries and Replies, and Art Travel Papers.

Monthly, price 6*d*. ; post free for a year, 7*s*. The yearly Volumes form a complete history of the Art World. 4to. cloth boards, 8*s*. 6*d*.

'*The only newspaper of the Art World published; and a very good one too.*'
JOURNALS AND JOURNALISM.

BAIRD.—*Works by the late Rev.* WILLIAM BAIRD, *M.A.*

THE DAYS THAT ARE PAST: A Manual of Early Church History. Fcap. 8vo. cloth boards, 2*s*. 6*d*.

THE INHERITANCE OF OUR FATHERS: Plain Words about the Book of Common Prayer. Fcap. 8vo. cloth boards, 3*s*. 6*d*.

WATCHING BY THE CROSS: Prayers, Readings, and Meditations for Holy Week. Royal 32mo. 6*d*.; cloth boards, red edges, 1*s*. [Fourth Edition.

BERTHA'S SCHOOL-FELLOWS. With Coloured Illustrations. 18mo. cloth boards, 1*s*. *A Collection of Short Stories from* '*SUNDAY.*'

BIRLEY.—WE ARE SEVEN. A Tale for Children. By CAROLINE BIRLEY. Coloured Illustrations by T. PYM. Square 16mo. extra cloth boards, 1*s*. 6*d*.

'*Delightfully quaint and full of life.*'—GUARDIAN.

BOODLE.—*Works by the Rev.* R. G. BOODLE, *M.A., Vicar of Cloford, Frome.*

THE LIFE AND LABOURS OF WILLIAM TYRRELL, D.D., First Bishop of Newcastle, New South Wales. With Portrait, two Maps, and Illustrations. Crown 8vo. cloth boards, 7*s*. 6*d*.

' npossible not to recognise in him a thoroughly devoted, single-hearted worker on the side of righteousness.'—PALL MALL GAZETTE.

WAYS OF OVERCOMING TEMPTATION: With a Form of Self-Examination and Prayers. Royal 32mo. paper cover, 6*d*.; cloth boards, 1*s*. [Ninth Edition.

BOURDALOUE.—EIGHT SERMONS FOR HOLY WEEK AND EASTER. Translated from the French of the Rev. FATHER LOUIS BOURDALOUE. By the Rev. G. F. CROWTHER, M.A., of St. John's College, Oxford. Crown 8vo. cloth boards, 3s. 6d.

BRAND.—THE LIFE OF WILLIAM ROLLINSON WHITTINGHAM, Fourth Bishop of Maryland. By W. FRANCIS BRAND. With Portrait and Fac-similes. 2 vols. imp. 8vo. cloth, bevelled boards, 24s.

BRETT.—LEGENDS AND MYTHS OF THE ABORIGINAL INDIANS OF BRITISH GUIANA. By the Rev. W. H. BRETT, B.D. Illustrated, crown 8vo. cloth boards, 3s. 6d.

BRIGHT THOUGHTS FOR THE MORNING: A Book of Simple Meditations for Young People. Printed on Toned Paper, with Frontispiece. Square 16mo. extra cloth boards, 1s.

These little Meditations are arranged so as to form Readings for a Month, and are intended as a Help to Young People in their Daily Devotions.

'*Fresh, hopeful, and to the point.*'—CHURCH BELLS.

BROOKS.—LECTURES ON PREACHING. By the Rev. PHILLIPS BROOKS, Rector of Trinity Church, Boston, U.S.A. Crown 8vo. cloth boards, 3s. 6d. [5th Thousand.

'*Nothing more really helpful has ever appeared.*'—NEW YORK CHURCHMAN.

BUCK.—JEM MORRISON THE FISHER-BOY, and THE TRIALS OF A VILLAGE ARTIST. By RUTH BUCK. With Coloured Illustrations. 18mo. cloth boards, 1s. 6d.

BULLEY.—GREAT BRITAIN FOR LITTLE BRITONS. By ELEANOR BULLEY. With numerous Illustrations of Places and People. Crown 8vo. cloth, bevelled boards, 3s. 6d. [3rd Thousand.

'*A very pleasant device for making geography agreeable.*'—GUARDIAN.

'*Great pains have been taken with this book, and it contains abundance of information.*'—SPECTATOR.

'*A capital story-book, and withal eminently instructive. Its tales are tales of the sea and land, of brave men and noble boys, of fox-hunting and whale-catching, of girl printers, and of the dinner feasts our princesses used to cook and serve when they were happy little girls at Osborne House.*'—NATIONAL CHURCH.

BOOKS PUBLISHED BY

BURROWS.—*Works by the Rev. H. W. BURROWS, B.D. Canon of Rochester.*

THE EVE OF ORDINATION. Fcap. 8vo. cloth limp, 1s. 6d.
[Third Edition.

LENTEN AND OTHER SERMONS. Fcap. 8vo. cloth boards, 2s. 6d. [Second Edition.

'They are striking, simple, brief, and impressive.'—CHRISTIAN WORLD.
'A brevity, born not of poverty, but of fulness.'—CHURCH TIMES.

CALTHROP.—*Works by the Rev. GORDON CALTHROP, M.A. Vicar of St. Augustine's, Highbury.*

THE BRAZEN SERPENT, and other Sermons, preached before the University of Cambridge. Crown 8vo. cloth boards, 3s.

MEMORIALS OF THE LIFE AND MINISTRY OF THE REV. W. B. MACKENZIE, M.A., late Vicar of St. James's, Holloway. With Portrait. Crown 8vo. cloth boards, 6s.
[Second Edition.

'His distinctiveness lay in the piety of his personal character, the devotedness of his pastoral consecration, and the simplicity, earnestness, and success of his preaching.'—BRITISH QUARTERLY REVIEW.

CARPENTER.—SHORT OUTLINE LESSONS FOR EACH SUNDAY IN THE CHRISTIAN YEAR. By the Right Rev. W. BOYD CARPENTER, D.D., Bishop of Ripon. 16mo. 6d.

CENTRAL AFRICA.—A Monthly Record of the Work of the Universities' Mission. 8vo. Monthly. 1d. Volumes, cloth boards, 2s.

CHATTERBOX.—Weekly, One Halfpenny; Monthly, in Wrapper, 3d. Annual Volumes, containing about Two Hundred Full-page Illustrations, Illustrated paper boards, cloth back, 3s.; extra cloth, bevelled boards, gilt edges, 5s. Cloth cases for binding, 1s. each.

A few copies of the following Volumes are still in print:—
3s. Edition—1872, 1875, 1876, 1878, 1881, 1882, 1883, 1884.
5s. Edition—1870, 1875, 1878, 1879, 1882, 1883, 1884.

The most popular children's magazine ever published. In a review of children's books the TIMES *says of the volume edition:* '*Chatterbox is one of the best children's books we have seen.*'

CHILD-NATURE.—By one of the Authors of 'Child-World.' Illustrated. Small square 16mo. cloth boards, gilt edges, 3s. 6d.

CHILD'S OWN STORY-BOOK.—In short Words and Large Type. With Coloured Plates by T. PYM. Square 16mo. extra cloth boards, 1s. 6d. [Second Edition.

CHURCH CONGRESS REPORTS.
BRIGHTON, 1874. 8vo. paper covers, 5s. 6d.; cloth boards, 6s. 6d.
STOKE-ON-TRENT, 1875. 8vo. paper covers, 5s. 6d.; cloth boards, 6s. 6d.
PLYMOUTH, 1876. 8vo. paper covers, 5s. 6d.; cloth boards, 6s. 6d.

CHORISTER'S ADMISSION CARD.—Contains Reasons and Motives for joining the Choir, with space for name, &c. On Card in Red and Black. 2d.

CLARKE.—COMMON-LIFE SERMONS. By the Rev. J. ERSKINE CLARKE, M.A., Vicar of Battersea, Hon. Canon of Winchester. Fcap. 8vo. cloth limp, 2s.; cloth boards, 2s. 6d. [6th Thousand.

CLARKE.—*Works edited by the Rev. J. ERSKINE CLARKE, M.A.*
CHILDREN OF THE OLD TESTAMENT. With numerous Full-page Illustrations. Royal 4to. paper bds., 1s. 6d.; cloth gilt, 2s. 6d.

CHILDREN'S HOME HYMN-BOOK. Royal 32mo. 1d.; cloth, 2d.

CHILDREN'S SCHOOL HYMN-BOOK. Royal 32mo. 1d.; cloth, 2d.

GOOD STORIES. The earlier numbers, consisting of 180 Complete Stories. Illustrated, in an Ornamental Cover, 3d. each. A List forwarded on application.

The following Volumes, strongly bound in cloth boards, 1s. 3d. each, will be found most useful in Village and Lending Libraries:—
ALICE AND HER CROSS, and Other Stories (Temperance).
COLONEL ROLFE'S STORY (Soldiers).
CONSULTING THE FATES, and Other Stories (Young Women).
FOUR LADS AND THEIR LIVES, and Other Stories (Confirmation).
GREGORY OF THE FORETOP, and Other Stories (Sailors).
JACK STEDMAN, and Other Stories (Young Men).
MARTIN GAY THE SINGER, and Other Stories (Temperance).
NETHER STONEY, and Other Stories (Temperance).
RHODA'S SECRET, and Other Stories (Young Women).
THE FORTUNE-TELLER, and Other Stories (Young Women)
THE RAINHILL FUNERAL, and Other Stories (Tradesmen).

CLARKE. — *Works edited by Rev. J. ERSKINE CLARKE.* — (*Continued.*)

THE PARISH LIBRARY. Illustrated. 18mo. cloth boards, price 1s. each.

DEB CLINTON, THE SMUGGLER'S DAUGHTER.
LUCY GRAHAM.
OLD ANDREW THE PEACEMAKER.
THE CLOCKMAKER OF ST. LAURENT.
CAN SHE KEEP A SECRET?

THE PARISH MAGAZINE. Illustrated, Monthly, 1d.; post free for a year, 1s. 6d.

The following Volumes are still to be had, and will be found very popular for School Libraries, Prizes, and for Lending to Sick Folk : —

1s. 6d. Edition— 1859, 1861, 1864, 1866, 1867, 1869, 1872, 1876, 1879, 1881, 1883, 1884.

2s. Edition— 1859, 1861, 1864, 1866, 1867, 1869, 1870, 1871, 1872 1873, 1874, 1877, 1879, 1881, 1883, 1884.

This was the first and the most popular Magazine adapted for localisation. Both the Archbishops, and nearly the whole of the Bishops, have from time to time contributed to its pages. 'Hints on Localising the Parish Magazine' will be forwarded on application.

THE PRIZE. For Boys and Girls. Monthly, 1d.; illustrated with numerous Engravings and one Coloured Picture. Post free for a year, 1s. 6d.

Each Volume contains about One Hundred Illustrations.

Volumes, 1s. 2d. paper cover; 1s. 6d. Illustrated paper boards; 2s. cloth boards; 2s. 6d. extra cloth boards, gilt edges.

Some copies of the following Volumes are still in print:—

1s. 2d. Edition— 1879, 1881, 1882, 1883, 1884.
1s. 6d. Edition— 1873, 1877, 1879, 1881, 1882, 1883, 1884.
2s. Edition—1874, 1879, 1881, 1882, 1883, 1884.
2s. 6d. Edition— 1877, 1879, 1880, 1881, 1882, 1883, 1884.

A New Series, with Coloured Illustrations, commenced with the January issue for 1882. The Volumes are most attractive, and contain Thirteen Coloured Plates and numerous Engravings.

'*A well-illustrated monthly serial of such literature as is calculated to please and benefit the younger boys and girls of our Sunday-schools. " The Children's Prize" is a meritorious and useful publication. For its special purpose—the reward and encouragement of industry and intelligence in the classes of schools for poor children—no better work lies upon our table.*'—THE ATHENÆUM.

COBB.—*Works by* JAMES F. COBB.

THE WATCHERS ON THE LONGSHIPS. A Tale of Cornwall in the Last Century. Illustrated by Davidson Knowles. Crown 8vo. cloth, bevelled boards, 3s. 6d. [Twelfth Edition.

'*A capital story, and one we heartily commend to boy readers, both gentle and simple.*'—GUARDIAN.

MARTIN THE SKIPPER. A Tale for Boys and Sea-faring Folk. Illustrated. Crown 8vo. cloth, bevelled boards, 3s. 6d.

'*We should imagine those queer folk indeed who could not read this story with eager interest and pleasure, be they boys or girls, young or old. We cannot sufficiently commend the style in which the book is written, and the religious spirit which pervades it.*'—CHRISTIAN WORLD.

OFF TO CALIFORNIA. A Tale of the Gold Country. Adapted from the Flemish. Illustrated by A. Forestier. Crown 8vo. cloth, bevelled boards, 3s. 6d.

CONVOCATION REPORTS.

THE SALE OF ADVOWSONS AND THE AUGMENTATION OF SMALL LIVINGS. Second Report of the Lower House of the Convocation of Canterbury, July 1879. 4d.

THE RELATIONS OF CHURCH AND STATE. A Full Report of the Committee of Convocation of Canterbury, July 1879. 4d.

THE RUBRICS OF THE BOOK OF COMMON PRAYER. The Report of Convocation of Canterbury, as presented to Her Majesty the Queen, in obedience to the Royal Letters of Business, on July 31, 1879. 1s.

CORAL MISSIONARY MAGAZINE.—A Record of Missionary Work among the Working Classes and in the Church Missionary Schools and Stations abroad. Monthly, 1d.; Post free for a year to all parts of the world, 1s. 6d.; Volumes, cloth, 1s. 6d. each; cloth cases for a year's numbers, 8d.

COWIE.—**THE VOICE OF GOD:** Chapters on Foreknowledge, Inspiration, and Prophecy. By the Very Rev. B. MORGAN COWIE, B.D., Dean of Exeter. Crown 8vo. cloth boards. 5s.

'*Not merely valuable, but indispensable, to the student of prophecy.*'
LITERARY CHURCHMAN.

CROMPTON.—A TALE OF THE CRUSADES. By SARAH CROMPTON. Fcap. 8vo. cloth boards, 1s.

Sir W. Scott's 'Talisman' in Short Words.

CUDDESDON MANUAL OF INTERCESSION FOR MISSIONS. 16mo. 4d. [Fourth Edition.

CUTTS.—THE BREAKING OF THE BREAD: An Explanation of the Holy Communion, with Notes on the Communion Service. By the Rev. E. L. CUTTS, B.A., D.D., Vicar of Holy Trinity, Haverstock Hill. 18mo. extra cloth boards, red edges, 2s.

THE DAISY: A Service of Song suitable for Sunday Schools and Temperance Societies. 8vo. 4d.

DANIEL.—*Works by the Rev. EVAN DANIEL, M.A., Principal of the National Society's Training College, Battersea; Hon. Canon of Rochester.*

THE PRAYER-BOOK: Its History, Language, and Contents. Crown 8vo. cloth boards, 6s. [Tenth Edition.

'We need say little more of Principal Evan Daniel's work than that its eighth edition is now before us. It is, if we rightly recollect, less than four years since we heartily commended the first edition to the notice of our readers. So large, wide, and speedy a circulation is proof at once of the interest which the subject possesses, and of the high merit of this treatise on it. . . . . The "glossarial notes" on the Prayer-book version of the Psalms are a peculiar and valuable ingredient in this serviceable volume; so also are the condensed, but pregnant, remarks upon the "Propria" for each of the Sundays and festivals, which will often furnish most valuable hints and references for the Sunday-school teacher and the preacher.'—GUARDIAN. Second Notice.

THE DAILY OFFICES AND LITANY. Being an Introduction to the Study of the Prayer-Book. Specially designed for the Use of National Schools and Sunday Schools. Fcap. 8vo. 8d.; cloth boards, 10d. [5th Thousand.

DE TEISSIER.—THE PARABLES OF OUR LORD JESUS CHRIST PRACTICALLY SET FORTH. By the Rev. G. F. DE TEISSIER, B.D., Rector of Church-Brampton. Fcap. 8vo. cloth limp, 2s. 6d.; cloth boards, 5s.

DICTIONARY OF THE ENGLISH CHURCH, ANCIENT AND MODERN. Crown 8vo. cloth boards, 7s. 6d.

'Besides containing much information, ecclesiastical and historical, is also of considerable practical utility. The writer is impartial and trustworthy.'
SPECTATOR.

DIVINE FELLOWSHIP.—A Daily Text-Book. 18mo. cloth boards, 9d.

DIX.—SERMONS, DOCTRINAL AND PRACTICAL. By the Rev. MORGAN DIX, Rector of Holy Trinity, New York. Crown 8vo. cloth boards, 3s. 6d.

DUMBLETON.—PRAYERS AND MEDITATIONS FOR THE MORNING AND EVENING OF EACH DAY OF THE WEEK. Chiefly in the Words of Holy Scripture. Arranged by the Rev. E. N. DUMBLETON, M.A., Rector of St. James's, Exeter. Fcap. 8vo. cloth boards, 1s. 6d.

FORMS OF PRAYER to accompany Sermons and Instructions, for use in Churches and Mission Rooms with the approval of the Ordinary. Crown 8vo. 9d.

DYER.—*Works by the Rev. A. SAUNDERS DYER, M.A.*

SKETCHES OF ENGLISH NONCONFORMITY. With Introductory Letter by the Bishop of Winchester. Crown 8vo. cloth, 1s. 6d.

'*We recommend the work to Churchmen and Nonconformists.*'
ECCLESIASTICAL GAZETTE.

OUR CLASS MEETING: A Bible-class Manual. 3d.

AN OFFICE FOR NEW-YEAR'S EVE. 1d.; 6s. per 100.

EDITH VERNON'S LIFE-WORK.—By the Author of 'Harry's Battles,' 'Susie's Flowers,' &c. &c. Crown 8vo. extra cloth boards, 3s. 6d.
[Tenth Edition.

'*A very pretty story, very well told. . . . . It is somewhat crowded with incident, but it is really well written; the lessons inculcated are excellent, and some of its characters are such as we shall not easily forget.*'—LITERARY CHURCHMAN.

EWING.—A WEEK SPENT IN A GLASS POND. By the Great Water-Beetle. Written by Mrs. EWING, Author of 'Six to Sixteen,' 'We and the World,' &c. With upwards of Forty Illustrations in Colours by R. ANDRÉ. 4to. Pictorial Cover, cloth back, paper boards, 3s. 6d.

'*A clever little fantasia on the keeping of an aquarium, by Mrs. J. H. Ewing, than whom a better qualified author on the subject could not have been found.*'—MORNING POST.

EDMUNDS.—SIXTY SERMONS: Adapted to the Sundays and Principal Holy-days of the Christian Year. By the Rev. JOHN EDMUNDS, M.A., formerly Fellow of the University of Durham. Fcap 8vo. cloth boards, 3s. 6d.

FAMILY WORSHIP FOR BUSY HOMES. On Folding Card, in plain type, 2*d*.

FARRAR.—THE CHRISTIAN MINISTRY: A Manual of Church Doctrine. By the Rev. THOMAS FARRAR, Rector of St. Paul's, Guiana. Crown 8vo. cloth boards, 6*s*. [Third and Enlarged Edition.

'Mr. Farrar's book is one which we would gladly see in the hands of all Readers, District Visitors, Teachers, and young men preparing for Holy Orders; for the selections have, for the most part, been made from writers who are both learned and clear.'—CHURCH TIMES.

FAVOURITE STORY-BOOK.— A Book for the Little Ones. Profusely Illustrated with Large Pictures, and Easy Reading. The Illustrations are printed in Sepia. Small 4to. paper covers, 1*s*.; cloth boards, 2*s*.

' A most attractive volume for juvenile readers. The stories would do very well to read out in school as exercises in composition. The book is handsome enough, however, to deserve a place on the drawing-room or parlour table, where even the older folk might dip into its contents with satisfaction.'
SCHOOLMASTER

FLYING LEAVES.— With Prefatory Note by the Rev. the EARL OF MULGRAVE. 32mo. fancy cloth boards, 9*d*.

' These little leaves are sent out into the world with the earnest prayer that one of them at least may be wafted to some weary soul, and carry with it one little ray of comfort and hope.'—INTRODUCTORY CHAPTER.

THE FIRST LADY OF THE LAND AND OTHER STORIES. Ten full-page coloured Illustrations. Crown 8vo. cloth, bevelled boards, 3*s*. 6*d*.

FOLLOWING CHRIST: Short Meditations for Busy People. Adapted from the French. 18mo. cloth limp, 1*s*.

' Thoroughly practical.'—SPECTATOR.

FOOTPRINTS: A Little Book of Choice Extracts. 64mo. fancy cloth boards, 9*d*.

FORDE.—THE OLD SHIP; or, Better than Strength. By H. A. FORDE With Full-page tinted Illustrations. Crown 8vo. cloth, bevelled boards, 3*s*. 6*d*.

FULTON.—*Works by the Rev.* JOHN FULTON, *D.D.*, *LL.D.*,
Rector of St. George's Church, St. Louis.

INDEX CANONUM. The Greek Text, an English Translation and Complete Digest of the entire Code of Canon Law of the undivided Primitive Church. Impl. 8vo. cloth boards, 10s. 6d.
[Second Edition, Revised and Enlarged.

THE LAWS OF MARRIAGE: Containing the Hebrew Law, the Roman Law, the Law of the New Testament, and the Canon Law of the Universal Church, concerning the Impediments of Marriage and the Dissolution of the Marriage Bond; Digested and Arranged with Notes Scholia. Crown 8vo. cloth boards, 7s. 6d.

GOOD STORIES.—New Series. Each with a Coloured Frontispiece, Monthly, 3d.; Series, containing four numbers, bound in extra cloth boards, 1s. 6d.; Five Volumes, extra cloth boards, gilt edges, 5s. each.

Most suitable for Prizes or Presents.

GOSPEL MISSIONARY.—Containing Missionary News, Anecdotes, and Verses suited for Young People. Illustrated. Monthly, ½d.

Published under the Direction of the S. P. G.

GRAIN OF MUSTARD SEED; or, Woman's Work in Foreign Parts. Monthly, 1d. Post free for a year to all parts of the world, 1s. 6d.

Published by the S.P.G. Ladies' Association for the Promotion of Female Education among the Heathen.

HARRIS.—GOLDEN STEPS: Lectures to Communicants' Classes. By the late Rev. G. C. HARRIS, M.A. Royal 32mo. 6d.; cloth boards, red edges, 1s. [Third Edition.

HAPPY SUNDAY AFTERNOONS.—A Series of Bible Outlines, printed on Superfine Lined Paper, for the Little Ones to colour and write about. Crown 4to. 1s.; cloth boards, 1s. 6d.

'*The publishers improve on the educational idea, which is at the root of the Kindergarten system, of making children teach themselves in their amusements. In "Happy Sunday Afternoons for the Little Ones" this firm, so pleasantly associated with a long history of juvenile recreation, supplies the means of self-improvement by a series of simple Bible outlines, to colour or write about either from memory or by reference to the Scriptures themselves.*'

THE DAILY TELEGRAPH.

HELEN MORTON'S TRIAL, AND TIMID LUCY.
With Coloured Illustrations. 18mo. extra cloth boards, 1s. 6d.

HELP AT HAND; or, 'What shall we do in Accidents or Illness?' By the COUNTESS COWPER. With numerous Illustrations. Fcap. 8vo. [*In the press.*

HELPS BY THE WAY,—
    I. MY MORNING HYMN. | III. MY WEEKLY QUESTIONS.
    II. MY DAILY RULES. | IV. MY CONFESSION TO GOD.
    Printed in red and black, 8vo. 1*d.* [Second Edition.

HER GREAT AMBITION.—A Story for Little Boys and Girls. With Thirty Illustrations. Small crown 4to. cloth, bevelled boards, 2*s.* 6*d.*

HIGH WAGES AND OTHER STORIES. Ten full-page coloured Illustrations. Crown 8vo. cloth, bevelled boards, 3*s.* 6*d.*

HINTS TO CHURCH-WORKERS.— Square 16mo. cloth boards, 1*s.* 4*d.*
    THE SAME AS TRACTS.
    HINTS TO CHOIRMEN, including an Office for the Admission of a Chorister. 1*d.* [Fifth Edition.
    HINTS TO SUNDAY-SCHOOL TEACHERS. 1*d.*
[Third Edition.
    HINTS TO LAY READERS, with the Form for Admitting Readers. 2*d.*
    HINTS TO TEACHERS OF ADULT CLASSES. 1*d.*
    HINTS TO LAY MISSIONERS. 1*d.*
    HINTS ON VISITING THE POOR AND SICK. 1*d.*
    HINTS TO BELL-RINGERS. 1*d.*
    Published under the Direction of the London Diocesan Lay-Helpers' Association.

HOBART.—*Works by the Hon. Mrs. C. HOBART, née N. P. W.*
    THE CHANGED CROSS. With Outline Illustrations by H. J. A. MILES. Square 16mo. cloth, bevelled boards, gilt edges, 1*s.* 6*d.*
[Twentieth Edition.
    THE CHANGED CROSS. Set to Music by GEORGE CARTER. 4*s.*
    THE CLOUD AND THE STAR. With Outline Illustrations by H. J. A. MILES. Square 16mo. cloth, bevelled boards, gilt edges, 1*s.* 6*d.*
[Second Edition.

HOBSON.— AIDS TO THE STUDY OF THE BOOKS OF SAMUEL. By the Rev. EDWIN HOBSON, M.A., Principal of St. Katharine's College, Tottenham. Fcap. 8vo. cloth boards, 2 vols. 1*s.* 6*d.* each, or 1 vol. complete, with Map, 2*s.* 6*d.*
    '*No difficulty is left unexplained, and the contents of the book are admirably summarised.*'—THE SCHOOLMASTER.

HOME REUNION SOCIETY'S PUBLICATIONS.

1. TWO PAPERS UPON THE RELATIONS OF THE ENGLISH CHURCH TO NONCONFORMITY. 3*d.*

2. I. THE CHURCH AND THE NONCONFORMISTS. By JOHN SHELLY. II. THE CHURCH IN RELATION TO HOME REUNION. By JOHN TREVARTHEN. III. THE CHURCH AND DISSENT. By Rev. CANON HOLE. 2*d.*

3. SERMON. By the LORD BISHOP OF WINCHESTER. 2*d.*

4. A LECTURE. By EARL NELSON. And Accounts of Two Meetings on Home Reunion at Salisbury. 3*d.*

5. ON THE PROMOTION OF A BETTER UNDERSTANDING BETWEEN CHURCHMEN AND NONCONFORMISTS: An Address at Ipswich Conference. By R. DENNY URLIN. Price 2*d.* [*Out of print.*

6. A PAPER ON PRIMITIVE EPISCOPACY. By EARL NELSON. 3*d.*

7. 'PEACE IN THE SACRAMENTS.' The New Congregational Hymn-book compared with the Book of Common Prayer. By the Rev. J. FOXLEY. 2*d.*

8. 'AN EIRENICON FOR THE WESLEYANS.' Two Prize Essays under the above title. By W. T. MOWBRAY and the Rev. V. G. BORRADAILE. 3*d.*

9. 'WHO IS RIGHT?' A Friendly Conversation between a Methodist and a Churchman. By the Rev. CANON BUCK, Rector of St. Dominick, Cornwall. 6*a.*

10. TWO PAPERS READ AT THE DIOCESAN CONFERENCE AT TRURO. By EARL NELSON and Rev. CANON BUSH. 2*d.*

11. THE DOCTRINE OF THE INCARNATION: Its Study the Key to the Knowledge of 'The whole Counsel of God,' and thereby to the Attainment of Reunion. By the Rev. H. C. POWELL, M.A., Rector of Wylye, Wilts. 4*d.*

12. PRIMITIVE CHURCH PRINCIPLES NOT INCONSISTENT WITH UNIVERSAL CHRISTIAN SYMPATHY. By the Rev. W. A. BUTLER, M.A. 2*d.*

HOLINESS TO THE LORD: The Character of the Christian Priest. Adapted from the French of the ABBÉ DUBOIS, for the use of the English Clergy. With an Introduction by the BISHOP OF CARLISLE. Crown 8vo. cloth boards, 7*s.* 6*d.*

BOOKS PUBLISHED BY

HONOR BRIGHT; or, The Four-Leaved Shamrock. By the Author of 'One of a Covey,' 'Robin and Linnet,' &c. With Full-page Tinted Illustrations. Crown 8vo. cloth, bevelled boards, 3s. 6d.
[Fourth Edition.

'*A cheery, sensible, and healthy tale.*'—TIMES.

HOPKINS.—*Works by the Rev. W. B. HOPKINS, B.D., Vicar of Littleport, Cambridge.*

HOLY SCRIPTURE: Temperance and Total Abstinence. Fcap. 8vo. cloth boards, 1s. [Third Thousand.

'*It will repay study either by the total abstainer or moderate drinker.*'
DAILY REVIEW.

THE POSITION AND DUTY OF NON-ABSTAINERS WITH REFERENCE TO THE TEMPERANCE CAUSE. 8vo. 4d. [Second Edition.

HOW.—*Works by the Right Rev. W. WALSHAM HOW, D.D., Bishop of Bedford and Suffragan of London.*

THE BOY HERO. A Story founded on Fact. Illustrated by H. J. A. MILES. Oblong, paper boards, 1s. 6d.

A PRAYER FOR THE PARISH. On Card, in red and black. 1d.

A SERVICE FOR THE ADMISSION OF A CHORISTER. In red and black. 2d.

CANTICLES POINTED FOR CHANTING, WITH APPROPRIATE CHANTS. 4to. paper covers, 1s.

DAILY FAMILY PRAYER. Fcap. 8vo. cloth boards, 1s. 6d.
[Twelfth Edition.

☞ A Sixpenny Edition, in large type, cloth boards, is now ready. This volume will be found most suitable for parochial distribution, and is the cheapest book of Family Prayers yet published.

HOLY COMMUNION. For those who need Encouragement. 6d. per Packet of Twenty. [153rd Thousand.

MORNING AND EVENING PRAYER FOR A CHILD. Cloth, 1d.

NOTES ON THE CHURCH SERVICE. Fcap. 8vo. cloth, 9d.

PASTORAL WORK. Fcap. 8vo. cloth boards, 2s. 6d.

'*There is not a dull page among the 150 which make up "Lectures on Pastoral Work," by the Bishop of Bedford. In six lectures, originally delivered to Divinity Students at Cambridge, Dr. Walsham How deals with the clergyman's "Equipment," his "Dangers and Difficulties," his duty of "Visitation," his "Dealing with Infidelity," his "Preaching," and the help he may get from the "Pastoral Epistles." Here is a pregnant quotation from the lecture on "Preaching"* . . . .'—GUARDIAN.

HOW.—*Works by Right Rev. W. WALSHAM HOW.*—(*Continued.*)

PASTOR IN PAROCHIÂ. Fcap. 8vo. cloth boards, 3*s.* 6*d.* With the Appendix. Cloth boards, red edges, 4*s.* 6*d.*; calf limp antique, 10*s.* 6*d.* Also morocco plain, and best flexible morocco, red under gold edges. [Fifteenth Edition.

PLAIN WORDS. First Series. Sixty Short Sermons for the Poor, and for Family Reading. Fcap. 8vo. cloth, turned in, 2*s.*; cloth boards, red edges, 2*s.* 6*d.* Large-type Edition, cloth boards, 3*s.* 6*d.*
[Forty-ninth Edition.

PLAIN WORDS. Second Series. Short Sermons for the Sundays and Chief Holy-days of the Christian Year. Fcap. 8vo. cloth, turned in, 2*s.*; cloth boards, red edges, 2*s.* 6*d.* Large-type Edition, cloth boards, 3*s.* 6*d.* [Thirtieth Edition.

Vols. I. and II., in one vol. cloth boards, 4*s.* 6*d.*

PLAIN WORDS. Third Series. Forty Meditations with a View to the Deepening of the Spiritual Life. Fcap. 8vo. cloth limp, 2*s.*; cloth boards, red edges, 2*s.* 6*d.* Large-type Edition, cloth boards, 3*s.* 6*d.*
[Seventeenth Edition.

PLAIN WORDS. Fourth Series. Forty Readings for those who desire to Pray Better. Fcap. 8vo. cloth limp, turned in, 2*s.*; cloth boards, red edges, 2*s.* 6*d.* [Seventh Edition.

Vols. III. and IV. in one, cloth boards, 4*s.* 6*d.*

PLAIN WORDS, as Tracts. Series I.-III., in Large Type, 2*s.* 6*d.* each Series.

A Selection from 'Plain Words;' for Parochial distribution, in smaller type, 1*s.* per packet: three kinds.

PLAIN WORDS TO CHILDREN. Crown 8vo. cloth, bevelled boards, 2*s.* 6*d.*; fcap. 8vo. cloth limp, turned in, 2*s.* [Third Edition.

PRAYERS FOR SCHOOLS. Royal 32mo. paper covers, 3*d.*; cloth, 6*d.*

PRIVATE LIFE AND MINISTRATIONS OF THE PARISH PRIEST. Royal 32mo. cloth, 6*d.*

RESOLUTIONS FOR THOSE RECOVERING FROM SICKNESS. On Card, in red and black, 12 copies in packet, 6*d.*

REVISION OF THE RUBRICS. An Historical Survey of all that has been done since the issue of the Ritual Commission in 1867, Demy 8vo. 1*s.*

SCRIPTURE READINGS. Selected Passages for Reading to the Sick. The Appendix to 'Pastor in Parochiâ.' Fcap. 8vo. cloth boards, 1*s.* 6*d.*

HOW.—*Works by Right Rev. W. WALSHAM HOW.*—(*Continued.*)

SEVEN LENTEN SERMONS ON PSALM LI. Fcap. 8vo. cloth limp, turned in, 1*s*. [Eleventh Edition.

SUGGESTIONS FOR OBSERVING THE DAY OF INTERCESSION. 1*d*.; 6*s*. per 100. [10th Thousand.

THE EVENING PSALTER POINTED FOR CHANTING. Oblong cloth limp, 6*d*.

TWENTY-FOUR PRACTICAL SERMONS. Fcap. 8vo. cloth limp, turned in, 2*s*.; cloth boards, red edges, 2*s*. 6*d*. [Twelfth Edition.

TWO ADDRESSES ON HOLY MARRIAGE. 1*d*.

VESTRY PRAYERS WITH A CHOIR. On Card, in red and black, 1*d*.

HOW.—**WEEK-DAY SERVICES IN COUNTRY CHURCHES.** By the Rev. F. DOUGLAS HOW, M.A., Rector of Evershot, Dorset. 6*d*.

HOW TO PRAY THE LORD'S PRAYER.—32mo. 1*d*.; 6*s*. per 100. [10th Thousand.

'*In the plainest type and language. We can earnestly commend it for distribution as likely to be most useful.*'—GUARDIAN.

INGELOW.—*Works by JEAN INGELOW.*

STUDIES FOR STORIES FROM GIRLS' LIVES. Illustrated. Crown 8vo. cloth boards, 3*s*. 6*d*. [Sixth Edition.

A SISTER'S BYE-HOURS. Illustrated. Crown 8vo. cloth boards, 3*s*. 6*d*. [Third Edition.

MOPSA THE FAIRY. Illustrated. Crown 8vo. cloth boards, 3*s*. 6*d*.

JONES.—*Works by C. A. JONES.*

COUNT UP THE SUNNY DAYS. A Story for Boys and Girls Illustrated. Small crown 8vo. cloth, bevelled boards, 3*s*. 6*d*.

FOUR LITTLE SIXES: A Story for Boys and Girls. Illustrated. Square 16mo. cloth boards, 1*s*. 6*d*.

ONLY A GIRL: A Story of a Quiet Life. A Tale of Brittany. Adapted from the French. With upwards of Forty Illustrations. Crown 8vo. cloth, bevelled boards, gilt edges, 3*s*. 6*d*.

'*We can thoroughly recommend this brightly written and homely narrative.*
SATURDAY REVIEW.

UNDER THE KING'S BANNER. Stories of the Soldiers of Christ in all Ages. With Introduction by the BISHOP OF BEDFORD Outline Illustrations by JOHN SADLER. Square 16mo. cloth boards, 2*s*. 6*d*.

KING.—ADDINGTON VENABLES, BISHOP OF NASSAU. A Sketch of his Life and Labours for the Church of God. By the Rev. W. F. H. KING, M.A., Commissary to the late Bishop. With Portrait, Map, and Illustrations. Crown 8vo. cloth, bevelled boards, 3s. 6d.

'*His life was one of utter self-denial and sheer hard work for God and His Church.*'—JOHN BULL.

KIP.—THE DOUBLE WITNESS OF THE CHURCH. By the Right Rev. W. INGRAHAM KIP, D.D., LL.D., Bishop of California. Crown 8vo. cloth boards, 3s. 6d.

This Edition of Bishop Kip's popular Lectures on the Principles of the Church is reprinted from the 22nd American edition, revised by the Author.

LAND OF LIGHT: A Transcript from the Rhythm of Bernard de Morlaix. With Outline Illustrations by H. J. A. MILES. Square 16mo. cloth, bevelled boards, gilt edges, 1s. 6d.

Uniform with 'The Changed Cross.'

LAY.—STUDIES IN THE CHURCH; Being Letters to an Old-Fashioned Layman. By the Right Rev. HENRY C. LAY, D.D., LL.D., Bishop of Easton, U.S.A. 18mo. cloth, bevelled boards, 2s. 6d.

'*We heartily commend this little book. . . . Full of clear sound common sense and manly piety, which cannot but enlist the reader's sympathy. The Bishop deals with his subject from an intellectual, doctrinal, and practical point of view. . . . Laymen cannot but appreciate his broad sympathies.*'—GUARDIAN.

LEA.—THE SUCCESSION OF EPISCOPAL JURISDICTION IN ENGLAND AT THE EPOCHS OF THE REFORMATION AND REVOLUTION. Exhibited in a Series of Tables. With an Introduction by JOHN WALTER LEA, B.A., F.G.S., &c., Fellow of the Royal Historical Society. Fcap. 4to. cloth boards, 5s.

'**LEFT TILL CALLED FOR.**'—By the Author of 'From Do-nothing Hall to Happy-Day House.' With Outline Illustrations by J. SADLER. Oblong, cloth boards, 1s.

The story of a little boy left at a railway station on Christmas Eve.

LEFT TO OUR FATHER: A Story for Children. By the Author of 'Clevedon Chimes,' &c. With Illustrations by H. FRENCH. Square 16mo. extra cloth boards, 1s. 6d.

LETTER OF COMMENDATION: A Card in Red and Black for giving to Parishioners on leaving a Parish. 6d. per packet of 12.

LEWIS.—YOUNG MEN'S BIBLE CLASSES AND HOW TO MANAGE THEM. By M. A. Lewis. Paper covers, 6d.

LITTLE FABLES FOR LITTLE FOLKS, which Great Ones may Read. With Nineteen Illustrations. 18mo. cloth boards, 1s.

LITTLE HELPS FOR DAILY TOILERS.—By a Working Associate of the Girls' Friendly Society. With Prefatory Note by the Bishop of Bedford. Royal 32mo. fancy cloth boards, 9d.

LITTLE LAYS FOR LITTLE LIPS.—With Outline Illustrations by H. J. A. Miles. Square 16mo. cloth, bevelled boards, gilt edges, 1s. 6d.
[Seventh Edition.

A LOST PIECE OF SILVER. By the Author of 'Edith Vernon's Life-Work,' &c. Illustrated. Crown 8vo. cloth boards extra, 3s. 6d.

'This is a simple, pathetic little story, which has the look of being true; true, that is, in the sense of being faithful to life. . . . Told without exaggeration, without any fine writing, but with very considerable power.'—Spectator.

LYTTELTON.—AIDS TO CHRISTIAN EDUCATION: Being a Brief Manual of Christian Doctrine and Practice. By the late Hon. and Rev. W. H. Lyttelton, M.A., Vicar of Hagley, Canon of Worcester. Vol. I. THE BAPTISMAL COVENANT. Fcap. 8vo. cloth boards, 2s. 6d.

This Volume is published in separate parts for Confirmation Candidates.

MACLAGAN.—WORDS OF COUNSEL ADDRESSED TO CONFIRMATION CANDIDATES ON THE EVE OF CONFIRMATION DAY. By the Right Rev. W. D. Maclagan, D.D., Bishop of Lichfield. Fcap. 8vo. 3d.

MACRITCHIE.—BY THE SEA OF GALILEE. A Poem. By Margaret S. MacRitchie. Tinted Outline Illustrations by H. J. A. Miles. Square 16mo. cloth, bevelled boards, gilt edges, 1s. 6d.

WELLS GARDNER, DARTON, AND CO. 19

MARRIAGE SERVICE.—Printed in Red and Black with Illustrations, and a Marriage Chorale. Bound in White and Gold. 6*d.*; or White Silk, 5*s.*

MAY.—THE CHRISTIAN COURSE; or, Helps to the Practice of Meditation. By the Rev. THOMAS MAY, M.A., Vicar of Leigh, Tunbridge. With Preface by the BISHOP OF BEDFORD. Royal 8vo. cloth boards, 5*s.* [Fourth Edition, Corrected and Enlarged.

MILES.—*Works Illustrated by H. J. A. MILES.*

FROM DO-NOTHING HALL TO HAPPY-DAY HOUSE. By the Author of 'Left till called for,' &c. Daintily printed in the best style of Chromo-lithography. Illustrated cover, cloth back, coloured edges, 1*s.* 6*d.*

'*A very pretty allegory for children. The illustrations are exceedingly chaste, and are most natural.*'—THE ARTIST.

OUTLINE PICTURES FOR LITTLE PAINTERS. 4to. in chromo-lithographic wrapper, 1*s.*; cloth, 1*s.* 6*d.* Printed in sepia on gray paper specially made for the purpose.

MISSION FIELD.—Containing a variety of Missionary Information, with a Record of the Proceedings of the S. P. G. Monthly, 2*d.* Post free for a year to all parts of the World, 3*s.*

MISSION LIFE.—Monthly Record of Home and Foreign Church Work. Monthly, 6*d.* Post free for a year to all parts of the World, 7*s.* Cloth cases for binding six months' numbers, 1*s.*

The following Volumes are still in print :—

1867-1870, reduced to 3*s.* 6*d.* each; 1871 (Part II. only), 1872-1884, 3*s.* 6*d.* each vol. of six months.

'*A well-edited repository of news from every part of the Mission field.*'
NONCONFORMIST.

MISSIONARY CONFERENCES.—REPORT OF THE MISSIONARY CONFERENCE HELD AT LONDON, 1875. Crown 8vo. paper, 2*s.* 6*d.*; cloth, 3*s.*

REPORT OF THE MISSIONARY CONFERENCE HELD AT OXFORD, 1877. Crown 8vo. paper, 2*s.* 6*d.*; cloth, 3*s.*

MISSIONARY PRAYERS FOR THE EXTENSION OF CHRIST'S CHURCH AT HOME AND ABROAD. For Private and Family Use. 18mo. paper covers, 6d.; cloth boards, 1s.

MITCHELL.—THE SUFFERER'S GUIDE. By ELIZABETH HARCOURT MITCHELL, Author of 'The Beautiful Face,' &c. Edited by the Rev. T. T. CARTER, M.A. Crown 8vo. cloth boards, 3s. 6d.

This Volume consists of Three Parts:—On Suffering in General—On Spiritual Sufferings—Suffering a Means of Perfection. [Second Edition.

MOBERLY.—SACRIFICE IN THE EUCHARIST: A Conversation. By the Rev. G. H. MOBERLY, M.A. Crown 8vo. 1s.

MONTH BY MONTH.—Poems for Children. With Twelve Illustrations by T. PYM. Sq. 16mo. cloth, bevelled boards, gilt edges, 1s. 6d.

'*We hardly know which to praise most, the quaint little illustrations or the poetry.*'—CHURCH TIMES.

MOORE.—SIMPLE GUIDE TO CHURCH DOCTRINE: Being an Explanation of the Church Catechism in Question and Answer, with Notes and Scripture Proofs. By BLANCHE MOORE. 16mo. 4d. [Third Edition.

MORNING STAR: Daily Texts for Little Children. Printed in red and black, 32mo. extra cloth boards, 9d.

The texts selected are specially simple and plain.

MOTHER'S UNION.—Containing Morning and Evening Prayers and Four simple Resolutions, in Red and Black. 1d.; 6s. per 100.

N. OR M.—By the Author of 'Honor Bright,' 'Peas-Blossom,' 'One of a Covey,' &c. With numerous Illustrations by H. J. A. MILES. Crown 8vo. cloth, bevelled boards, 3s. 6d.

NORTON.—*Works by the Rev. J. G. NORTON, M.A., Rector of Christ Church Cathedral, Montreal.*

HEARTY SERVICES: or, Revived Church Worship. Crown 8vo. cloth, bevelled boards, 3s. 6d. [Third Edition.

WORSHIP IN HEAVEN AND ON EARTH: Responsive, Congregational, Reverent, Musical, and Beautiful. Demy 8vo. cloth boards, 12s. 6d.

O'BRIEN.—*Stories by Mrs. CHARLOTTE O'BRIEN.*

>**MARGARET AND HER FRIENDS.** With Coloured Frontispiece. Fcap. 8vo. cloth boards, 1*s*.
>
>**MOTHER'S WARM SHAWL.** With Coloured Frontispiece. Fcap. 8vo. cloth boards, 1*s*.
>
>**OLIVER DALE'S DECISION.** With Coloured Frontispiece. Fcap. 8vo. cloth boards, 1*s*.

ONE OF A COVEY.—By the Author of 'Honor Bright,' 'Peas-Blossom,' &c. With numerous Illustrations by H. J. A. MILES. Crown 8vo. extra cloth boards, gilt edges, 3*s*. 6*d*.

> '*Full of spirit and life, so well sustained throughout that grown-up readers may enjoy it as much as children. . . . It is one of the best books of the season.*'
> GUARDIAN.
>
> '*One of the chief characters would not have disgraced Dickens' pen.*'
> LITERARY WORLD.

THE OLD, OLD STORY. By the Author of 'Heart to Heart,' &c. With Outline Illustrations by H. J. A. MILES. Square 16mo. cloth, bevelled boards, gilt edges, 1*s*. 6*d*.

O'REILLY.—*Works by Mrs. ROBERT O'REILLY.*

>**CHILDREN OF THE CHURCH:** or, Short Lessons on the Church Catechism for Infant Children. 18mo. cloth boards, 1*s*. 6*d*. [Eighth Edition.
>
>**CHILDREN OF THE CHURCH.** Second Series. Lessons on the Collects. 18mo. cloth boards, 1*s*. 6*d*.
>
>**STORIES THEY TELL ME;** or, Sue and I. With Illustrations by H. J. A. MILES. Crown 8vo. cloth, bevelled boards, 3*s*. 6*d*.
>
>'*A thoroughly delightful book, full of sound wisdom as well as fun.*'
> ATHENÆUM.

OUR BOYS AND GIRLS BOTH GOOD AND BAD.—
Edited by the Author of 'Great Britain for Little Britons.' With Thirty Illustrations. Small crown 8vo. cloth, bevelled boards, 2*s*. 6*d*.

OUR WAIFS AND STRAYS.—
The Monthly Record of the Work of the Church of England Central Home for Waifs and Strays. 1*d*.

PALMER.—TRUE UNDER TRIAL. By FRANCES PALMER. With Tinted Illustrations from Drawings by G. L. SEYMOUR and W. A. CRANSTON. Crown 8vo. cloth, bevelled boards, 3s. 6d. [Fourth Edition.

'One of the best boys' books we have seen for a long time. Written with a rare combination of religious spirit, with a perfect abstention from cant; and so well put together, that we believe no reader who once took up the book would put it down without finishing it.'—STANDARD.

'A well-written story.'—TIMES.

DOGGED JACK. With Full-page Coloured Illustrations. Crown 8vo. cloth, bevelled boards, 3s. 6d. [Second Edition.

PAPAL CLAIMS CONSIDERED IN THE LIGHT OF SCRIPTURE AND HISTORY. With an Introduction by the BISHOP OF BEDFORD. Fcap. 8vo. cloth boards, 2s.

PARABLES OF THE KINGDOM.—Our Lord's Parables Simply Told for Children. By the Author of 'Voices of Nature,' 'Earth's Many Voices.' With 46 Illustrations in Outline by H. J. A. MILES. Square 16mo. cloth, bevelled boards, 2s. 6d.

'Teachers will find this a valuable help.'—CHURCH BELLS.

PEAS-BLOSSOM.—By the Author of 'Honor Bright,' 'One of a Covey,' &c. With numerous Illustrations by H. J. A. MILES. Crown 8vo. cloth boards, gilt edges, 3s. 6d.

'A delightfully written book for boys about twelve. The best book of the season.'—STANDARD.

PLAIN TEXTS FOR DAILY USE.—With Introduction by the BISHOP OF BEDFORD. 64mo. cloth boards, 4d.; Persian limp, 9d.; calf limp, 1s. 6d.

POTTER.—A PRESENT CHRIST: Daily Ten Minutes' Readings for Four Weeks on the Incarnation. By the Rev. J. HASLOCH POTTER, M.A., late Editorial Secretary to the Church of England Temperance Society. Fcap. 8vo. cloth boards, 2s. 6d. [Second Edition.

Specially suited for use at Daily Services or Family Prayer.

POTTER.—SERMONS OF THE CITY. By the Rev. HENRY C. POTTER, D.D., Secretary to the House of Bishops, U.S.A. Crown 8vo. cloth boards, 3s. 6d.

PRESCOTT.—*Works by the Rev. G. F. PRESCOTT, M.A., Vicar of St. Michael's, Paddington.*

    COUNSELS ON PRAYER. Royal 32mo. 6d.; cloth boards, 1s.

    HINDRANCES TO SPIRITUAL LIFE: A Course of Lent Lectures. Crown 8vo. cloth boards, 1s. 6d. [Fourth Edition.

    LECTURES ON THE LORD'S PRAYER. Crown 8vo. cloth boards, 1s. 6d.

THE 'PRIZE' BIBLE.—Numerous Illustrations. Twelve Coloured Chromo Plates. Large-type Letterpress, and printed on Toned Paper. 4to. extra cloth, bevelled boards, gilt edges, 12s. 6d.

PYM.—*Illustrated Works by T. PYM.*

    OUTLINE ILLUSTRATIONS FOR THE LITTLE ONES TO COLOUR. Fcap. 4to. 1s.; cloth, 1s. 6d.

    MORE OUTLINES for the LITTLE ONES TO COLOUR. Fcap. 4to. 1s.; cloth, 1s. 6d.
        These Outline Picture-books are printed in sepia, on tinted paper expressly made for the purpose.

CHILDREN BUSY,
      CHILDREN GLAD,
CHILDREN NAUGHTY,
      CHILDREN SAD.

With Stories by L. C. An Illustrated Book of Child-life, printed in the best style of Chromo-lithography. Bound in an Illustrated cover, paper boards, cloth back, 3s. 6d.

    '*A most charming book for children. The pictures are very pretty, and the children represented in them look like real children as they are seen in nurseries, which is not always, nor even often, the case in books of this sort. Good writing also is not abundant in literature for the young, and for this reason the excellent stories by L. C. will be the more attractive. They display a delicate fancy, and will be read with real pleasure for their literary merit by grown-up people as well as children.*'—TIMES.

PYM.—*Illustrated Works by T. PYM.—(Continued.)*

PICTURES FROM THE POETS. A Book of Selected Extracts from Ancient and Modern Sources, illustrated by Child-life, carefully printed in Chromo-lithography in Brown and Red Tints. Oblong 4to. Illustrated cover, cloth back, 3s. 6d.

'T. Pym has fully established his reputation. . . . A succession of charming illustrations, admirably produced.'—NONCONFORMIST.

'This volume will afford pleasure to every one who can admire good artistic work.'—SCOTSMAN.

THE SNOW QUEEN. By HANS CHRISTIAN ANDERSEN. Beautifully printed in the best style of Chromo-lithography. The Illustrations represent Gerda and Kay's adventures in various countries, as contained in one of the most popular of Hans Andersen's stories. Fcap. 4to. Illustrated cover, cloth back, coloured edges, 5s.

THE QUEEN'S SHILLING AND OTHER STORIES.—With Ten full-page coloured Illustrations. Crown 8vo. cloth, bevelled boards, 3s. 6d.

THE QUIET HELPER. Text, Prayer, and Hymn for Four Weeks, and for the principal Holy-days. Printed in large type for hanging on the wall. On roller, 3s.

READINGS AND DEVOTIONS FOR MOTHERS. With Introduction by the BISHOP OF SALISBURY. Fcap. 8vo. cloth limp, 1s. 6d.

'This little book certainly seems to supply a want which has not exactly been supplied before, and as the Bishop of Salisbury observes, it has done it in a way most likely to prove useful.'—CHURCH TIMES.

ROBIN AND LINNET. By the Author of 'Honor Bright,' &c. With Coloured Illustrations by T. PYM. Square 16mo. extra cloth boards, 1s. 6d.

'The adventures are very amusing, and the story will be a favourite with children, who will delight to fancy themselves roaming on the beach, or dabbling in the pools in such liberty.'—GUARDIAN.

ROCHESTER DIOCESAN DIRECTORY. With Coloured Map of England and Wales, divided into the Dioceses. Published by Authority. Small crown 8vo. paper boards, cloth back, 1s. 6d.

WELLS GARDNER, DARTON, AND CO. 25

A ROUGH DIAMOND AND OTHER STORIES.—Ten full-page coloured Illustrations. Crown 8vo. cloth, bevelled boards, 3s. 6d.

ROWLEY.—*Works by the Rev. Henry Rowley.*

    THE RELIGIONS OF THE AFRICANS. Fcap. 8vo. cloth boards, 3s. 6d.

    TWENTY YEARS IN CENTRAL AFRICA. The Story of the Universities' Mission, from its Commencement under Bishop Mackenzie to the Present Time. With Map. Crown 8vo. cloth boards, 3s. 6d.

    '*The volume abounds in thrilling incidents.*'—American Churchman.

RUTH HALLIDAY; or, The Adopted Daughter. A Tale founded on Fact. Fcap. 8vo. cloth boards, 1s.

ST. AUSTIN'S COURT; or, The Grandchildren. With Coloured Frontispiece. 18mo. cloth boards, 1s.

SHADOWS OF TRUTH; or, Thoughts and Allegories in Prose and Verse. By G. M. C. Fcap. 8vo. cloth boards, gilt edges, 2s. 6d.
[Third Edition.

SHERLOCK.—THE AMETHYST: A Selection of Temperance Readings in Prose and Verse. By Frederick Sherlock. Crown 8vo. cloth boards, 1s.

SIDEBOTHAM.—THE DISCIPLINE OF TEMPTATION, and other Sermons. By the Rev. H. Sidebotham, M.A., Canon of Gibraltar and Chaplain of St. John's, Mentone. Fcap. 8vo. cloth boards, 1s.

SIKES.—*Works by the Rev. T. B. Sikes, M.A., F. R. His. Soc., Rector of Burstow, Surrey.*

    ENGLAND'S PRAYER-BOOK. A Short and Practical Exposition of the Services. 18mo. cloth boards, 2s. 6d.

    HISTORY OF THE CHRISTIAN CHURCH FROM THE FIRST TO THE SIXTEENTH CENTURY. 18mo. cloth boards, 3s. 6d.

    THE HOLY COMMUNION. Cloth, 6d.

BOOKS PUBLISHED BY

SILVERMERE ANNALS; Tales of Village Life. By C. E. B. With Coloured Frontispiece. Fcap. 8vo. cloth boards, 1s.

SISTER LOUISE. The Story of her Life-Work. With Portrait. Fcap. 8vo. cloth boards, 2s. 6d. *The Sister Dora of America.*

SKEY.—DOLLY'S OWN STORY. Told in her Own Words. By L. C. SKEY. Illustrated in Outline by J. SADLER. Square 16mo, cloth boards, 1s. 6d.

SLATTER.—STUDENTS' GOSPEL HARMONY. Being the Four Gospels in the Original Greek, arranged in parallel columns so as to show the consentient portions together, with a Preface and Analytical Tables. By the Rev. JOHN SLATTER, M.A., Vicar of Streatley, Berks, and Hon. Canon, Christ Church, Oxford. Demy 8vo. cloth boards, 12s. 6d.

> '*Although, in many cases, the assignation of a passage to a particular place in the "Harmony" must involve a choice of difficulties, there is evidence throughout that all the conditions of the problem have been fully considered. In short, a student who follows Canon Slatter will hardly go wrong.*'—DAILY NEWS.

SNOWDON.—*Works by* JASPER W. SNOWDON.

ROPE-SIGHT: An Introduction to the Art of Change-Ringing. Crown 8vo. paper cover, 1s. 6d. [Second Edition.

> '*Very much wanted. . . . Mr. Snowdon is peculiarly fitted for the post of teacher.*'—CHURCH BELLS.

STANDARD METHODS in the ART of CHANGE-RINGING. With a Book of Coloured Diagrams. Crown 8vo. paper covers, 2s. 6d.

A TREATISE ON TREBLE BOB. Crown 8vo. paper cover.—Part I., 1s.; Part II., 2s.

Part I.—History of Treble Bob—The In-and-Out-of-Courses of the Changes—Pricking Touches and Peals—Qualities of Peals—Transposition of Peals—Proof of Treble Bob—Lockwood's System—Conducting and Calling Round.

Part II.—Collection of 1000 Peals—Five-part Peals—Two-part Peals—One-part Peals—Peals with Tenors parted—Treble Bob Royal and Maximus—Composers' Names.

SONGS AND LYRICS FOR LITTLE LIPS. With Musical Contributions by W. H. CUMMINGS and others. Illustrated by G. L. SEYMOUR and others. 8vo. cloth extra, bevelled boards, gilt edges, 6s.

> '*A collection of some of the choicest little poems for children that we possess— some old, some new—with appropriate music and charming "pictures." We can imagine no more delightful present to a child six or seven years old.*'
> GUARDIAN.

SPECIAL SERVICES.

SPECIAL SERVICE OF INTERCESSION FOR THOSE AT SEA. 1*d.*; 6*s.* per 100.

Approved by the late Archbishop of Canterbury.

A SPECIAL SERVICE OF PREPARATION FOR HOLY COMMUNION. 50 Copies for 2*s.*

Many churches have adopted this Service for use on a week-day evening. It is now published with the view of an extended circulation, and is sanctioned by the Bishops of Winchester, Exeter, and others.

FORM OF INSTITUTION AND INDUCTION. 3*s.* per 100.

As prepared by a Committee of the Lower House of the Convocation of Canterbury.

FORM OF SERVICE FOR THE DEDICATION OF CHURCH BELLS. 2*s.* 6*d.* a packet of 50 copies.

AN OFFICE FOR NEW-YEAR'S EVE. 1*d.*; 6*s.* per 100 copies.

Approved by the Bishop of Winchester.

STEVENS.—LOVE IS OF GOD, and other Sermons. By the Right Rev. W. BACON STEVENS, D.D. LL.D., Bishop of Pennsylvania. Crown 8vo. cloth boards, 3*s.* 6*d.*

STONE. — *Works by the Rev. S. J. STONE, M.A., Vicar of St. Paul's, Haggerstone.*

DEARE CHILDE. A Village Idyl. With Outline Illustrations by H. J. A. MILES. Square 16mo. cloth, bevelled boards, gilt edges, 1*s.* 6*d.*
[Third Edition.

'*We scarcely know whether Mr. Stone—whose devotional poetry needs n praise—or Miss Mites has most happily caught the spirit that should animate such a memorial. Both are so true as to be almost painful—at least, to those who have loved and lost a little child.*'—NONCONFORMIST.

HYMNS FOR THE DAY OF INTERCESSION. 2*s.* 6*d.* per 100.
[80th Thousand

STORIES & EPISODES OF HOME MISSION-WORK.

With a Preface by the late ARCHBISHOP OF CANTERBURY Crown 8vo cloth boards, 3*s.* 6*d.*

STRACHAN.—FROM EAST TO WEST; or, Glance at the Church's Work in Distant Lands. By the Right Rev. J. M STRACHAN, M.D., D.D., Bishop of Rangoon. Illustrated crown 8vo. cloth boards, 3*s.* 6*d.*

SUNDAY.—Weekly, One Halfpenny; Monthly, in Wrapper, 3*d.* '
Annual Volumes, with upwards of Two Hundred Illustrations, illustrated paper boards, cloth back, 3*s.*; cloth, bevelled boards, gilt edges, 5*s.*

A few copies of the following Volumes are still to be had, in extra cloth, bevelled boards, gilt edges, price 5*s.* each:—1878, 1879, 1881, 1882, 1883, 1884.

'*If this Magazine finds its way to the glance of a child, woe be to the parent or friend standing by who does not chance to have the disposition or the means for purchasing it. It is absolutely full of charming pictures and interesting reading. The pictures are unquestionably better than those which one finds in some similar periodicals.*'—CHRISTIAN WORLD.

'*We know of no better magazine of its kind, and we imagine no handsomer gift at Christmas time to bestow upon a child.*'—CHURCH TIMES.

SWAYNE.—THE MINISTER OF CHRIST IN THESE LAST DAYS. Five Addresses to Candidates for Ordination. By the Rev. ROBERT S. SWAYNE, M.A., Chancellor and Canon Residentiary of Salisbury Cathedral. Crown 8vo. cloth boards, 2*s.*

TAYLOR.—OUT OF THE WAY. A Village Temperance Story. By H. L. TAYLOR. With numerous Illustrations by A. H. Collins. Crown 8vo. extra cloth, boards, gilt edges, 3*s.* 6*d.*

'*A favourable specimen. . . . We were half way through the book before we found out that it was a temperance tale at all.*'—GUARDIAN.

'*The book deserves warm praise; we wish there were more temperance tales like it.*'—CHRISTIAN WORKD.

'*This is a very readable and unusually well-written tale.*'—CHURCH BELLS.

TEMPERANCE HYMNS AND SONGS. With Accompanying Tunes. To which is prefixed a short Opening Service. Demy 8vo. paper covers, 1*s.* 6*d.*; cloth boards, 2*s.* 6*d.* [New and Enlarged Edition.

Published under the Direction of the Church of England Temperance Society.

This is the most comprehensive and the cheapest book of its kind yet issued.

THORNE.—*Works by E. H.* THORNE, *late Organist of Chichester Cathedral.*

A SELECTION OF SINGLE AND DOUBLE CHANTS. Oblong, cloth limp, 1*s.* [Twenty-third Edition.

PSALTER and CANTICLES, Pointed for Chanting to Anglican Chants, with Accents and Marks of Expression. 18mo. cloth limp, 1*s.*

THOUGHTS FOR THE SICK AND INFIRM ON THE DAY OF INTERCESSION. With Commendation by the BISHOP OF BEDFORD. 1*d.*; 6*s.* per 100.

THREE CUPS; or, The Girls of St. Andrews. By the Author of 'Marty and the Mite-Boxes.' Illustrated. Crown 8vo. cloth boards, 3s. 6d.

> 'It appears there is a Chinese proverb, "When you have three cups to drink, drink your three cups," and this gives the quaint title of "The Three Cups" to a pleasant little American book.'—GUARDIAN.

TITCOMB.—PERSONAL RECOLLECTIONS OF BRITISH BURMA. By Right Rev. J. H. TITCOMB, D.D., First Bishop of Rangoon. Illustrated. Demy 8vo. half cloth, 2s. 6d.

TOPSEY TURVY. By ALLEBYS. With numerous Illustrations by H. J. A. MILES. 4to. extra cloth boards. [*In the press.*

TROYTE.—CHANGE-RINGING. An Introduction to the Early Stages of the Art of Church or Handbell Ringing, for the Use of Beginners. By CHARLES A. W. TROYTE, of Huntsham Court, Devonshire; Member of the Ancient Society of College Youths, London. Crown 8vo. paper covers, up to 'Six Bells,' 1s.; Complete Edition, cloth limp, 2s. 6d.

TROYTE.—THE CHANGE-RINGERS' GUIDE TO THE STEEPLES OF ENGLAND. With an Appendix, containing information on many subjects interesting to the Exercise. Compiled by Rev. R. ACLAND-TROYTE and J. E. ACLAND-TROYTE, Members of the Ancient Society of College Youths, the Oxford University Society, &c. Cr. 8vo. 1s. 6d.

TUCKER.—*Works by the Rev. H. W. TUCKER, M.A., Secretary to the Society for the Propagation of the Gospel.*

MEMOIR of the LIFE and EPISCOPATE of EDWARD FEILD, D.D., Bishop of Newfoundland, 1844-1876. With Prefatory Note to the Author by the Right Hon. W. E. GLADSTONE, M.P. With Map of Newfoundland. Cr. 8vo. cloth boards, 5s. [Third Edit.

'*A record of a devoted life.*'—PALL MALL GAZETTE.

MEMOIR of the LIFE and EPISCOPATE of GEORGE AUGUSTUS SELWYN, D.D., Bishop of New Zealand, 1841-1869; Bishop of Lichfield, 1867-1878. With Two Portraits, Map, Facsimile Letters, and Engraving of the Lady Chapel of Lichfield Cathedral. Two vols. 8vo. 800 pp., cloth boards, 24s.

Popular Edition, 2 vols. crown 8vo. cloth boards, 12s.

> '*Of more than ordinary interest. . . . We shall rejoice if our endeavour to depict him shall send many to the pages of his Biography for fuller information.*'
> QUARTERLY REVIEW.

TUCKWELL.—THE MAGDALEN PSALTER. The Psalms, Canticles, and the Athanasian Creed, pointed for Chanting. By the Rev. L. S. TUCKWELL, M.A., Rector of Standlake, and J. STAINER, M.A., Mus.Doc., Organist of St. Paul's Cathedral. 18mo. cloth boards, 1s. 6d. [Fifth Edition.

TUTTIETT.—*Works by the Rev.* LAWRENCE TUTTIETT.
    COUNSELS OF A GODFATHER. Fcap. 8vo. cloth, bevelled boards, 2s. 6d. *Or in Five Parts.* [Second Edition.
    HOUSEHOLD PRAYERS FOR WORKING MEN. 18mo. cloth, 6d. [10th Thousand.
    PLAIN FORMS OF HOUSEHOLD PRAYER for Four Weeks. Chiefly for those Engaged in Necessary Business. In large type, fcap. 8vo. cloth boards, 2s. 6d. [Fifth Edition.
    THE TRUE PENITENT: Reflections on the Penitential Psalms. Fcap. 8vo. cloth, 1s. 6d.

TURNING-POINT OF LIFE, and **THE DOUBLE WARFARE:** Two Confirmation Stories. Illustrated. 18mo. cloth boards, gilt edges, 1s.

UNDER MOTHER'S WING. By L. C., Author of the Stories in 'Children Busy.' With Coloured Illustrations on every page by J. K. 4to. Illustrated Cover, 4s.

VENABLES.—*Works by the Rev.* GEORGE VENABLES, *S.C.L. Vicar of Great Yarmouth.*
    CHURCHMAN'S MANUAL. 32mo. cloth boards, 1s. [Third Edition. Enlarged.
    FIVE OFFICES FOR PAROCHIAL USE.—Sunday-school Teachers, Opening and Closing Schools, Meeting of District Visitors, Mission-room Services, Cottages, &c. Fcap. 8vo. 6d.
    OFFICES OF HOLY BAPTISM, CONFIRMATION, SOLEMNIZATION OF MATRIMONY, AND BURIAL OF THE DEAD, with Explanatory Observations. Royal 16mo. 3d. each. Or the Four in one vol. paper cover, 1s.; cloth, 2s.
    OUR CHURCH AND OUR COUNTRY. Crown 8vo. paper covers, 6d. [Sixth Edition.
    TEACHING OF JESUS CHRIST UPON MANY SUBJECTS OF THE CHRISTIAN VERITY. Fcap. 8vo. paper, 1s.; cloth limp, 2s. 6d.
    THREE EXTRA SERVICES FOR USE IN CHURCH— an Office for Communicants, a Service of Song, and an Office for a Catechetical Service. 24mo. 4d.

VERNON.—KALENDAR NOTES: Short Devotional Comments for Every Sunday and Holy-day of the Christian Year. By the Rev. J. R. VERNON, M.A., Rector of St. Audrie's, Somerset. Fcap. 8vo. cloth boards, 3s.
    '*Brief readings—terse and thoughtful.*'—LITERARY CHURCHMAN.

VIDAL.—*Stories by Mrs.* VIDAL.
    LUCY HELMORE. Coloured Illustrations. Fcap. 8vo. extra cloth boards, 2s. 6d.
    THE TRIALS OF RACHEL CHARLCOTE. With Coloured Frontispiece. Fcap. 8vo. cloth boards, 1s.

VOICES OF NATURE. By the Author of 'Parables of the Kingdom,' 'Earth's Many Voices,' &c. With Illustrations. Square 16mo. extra cloth boards, 1s. 6d.

WAYNE.—OLD PATHS: Sermons on the Apostles' Creed. By the Rev. E. F. WAYNE, M.A. Fcap. 8vo. cloth boards, 3s.

WHITWORTH.—*Works by the Rev. W. ALLEN WHITWORTH, M.A., Fellow of St. John's College, Cambridge.*

A BIBLE-CLASS MANUAL—Offices, Prayers, and Intercessions for Members of Bible Classes. Royal 32mo. 4d.; extra cloth, red edges, 6d.

THE DIVINE SERVICE. Holy Communion according to the Use of the Church of England, with Explanatory Notes and Helps to Private Devotion. Fcap. 8vo. paper boards, 8d.; cloth boards, 1s.
[Second Edition.

The same book, in cloth boards, bound with MY PRIVATE PRAYER-BOOK, containing spaces for notes of Special Intercession, 1s. 6d. complete.

'*The office is printed in larger type at the top, and the notes and helps in smaller at the bottom of each page. There is careful instruction on preparation and thanksgiving. A book we can recommend.*'—LITERARY CHURCHMAN.

IS IT PEACE? Words of Encouragement for Anxious Souls. 16mo. 6d.; extra cloth, 1s.

'*A rousing little book.*'—GUARDIAN.

MY PRIVATE PRAYER-BOOK. Containing spaces for notes of Special Intercession. Fcap. 8vo. 3d.

SEVEN PRAYERS ON THE SEVEN WORDS FROM THE CROSS. For Distribution on Good Friday, or for use during 'The Three Hours.' Fifty copies in packet, 1s. 9d.

THE CHURCHMAN'S ALMANAC FOR EIGHT CENTURIES (1201 to 2000), giving the Name and Date of every Sunday. Small folio, 2s. 6d.

THE NECESSITY OF PERSONAL TESTIMONY IN SEEKING THE CONVERSION OF THE UNGODLY: and A Plea for Open Prayer-Meetings. Two Addresses on Home Mission Work. Demy 8vo. 6d.

WILBERFORCE.—SPEECHES ON MISSIONS. By the Right Rev. SAMUEL WILBERFORCE, D.D. Edited by the Rev. HENRY ROWLEY. 8vo. cloth boards, 7s.
[Second Edition.

'*This book may be considered an instruction-book, not only by its provision of so much material ready to hand, but the lessons it gives as to the real importance and stirring interest of many facts and figures which seem uninteresting enough as generally read in missionary periodicals. It teaches the speaker or the preacher how to provide himself well with materials, and then how to infuse life and warmth into them.*'—CHURCH BELLS.

WILKINSON.—*Works by the Right Rev. G. H. WILKINSON, D.D. Bishop of Truro.*

ABSOLUTION : A Sermon. 8vo. paper covers, 1s.

BE YE RECONCILED TO GOD. 1d.; 6s. per 100. [New Edit.

CONFESSION : A Sermon. 8vo. paper covers, 1s.

FIRST STEPS TO HOLY COMMUNION. The Substance of Four Simple Instructions after Confirmation. Fcap. 8vo. 6d. A Superior Edition in Old Style, bound in white, 1s.

HINDRANCES and HELPS to the DEEPENING of the SPIRITUAL LIFE AMONG CLERGY and PEOPLE. 3d.

HOLY WEEK AND EASTER. Fcap. 8vo. cloth boards, 1s.
[Fifth Edition.

HOW TO KEEP LENT. Notes on Quinquagesima Sunday Address. 3d. [11th Thousand.

HOW TO DEAL WITH TEMPTATION. A Lenten Address. Fcap. 8vo. 3d.

INSTRUCTIONS IN THE DEVOTIONAL LIFE. Fcap. 8vo. price 6d. A superior edition, 1s. [52nd Thousand.

INSTRUCTIONS IN THE WAY OF SALVATION. Fcap. 8vo. price 6d. A superior edition, 1s. [26th Thousand.

LENT LECTURES. Fcap. 8vo. cloth boards, 1s. [15th Thousand.

MORNING AND EVENING PRAYERS FOR CHILDREN. On Card, 1d.

PENITENTIARY WORK: Its Principles, Method, Difficulties, and Encouragements. Fcap. 8vo. price 6d.

PRAYERS FOR CHILDREN. 32mo. 2d.

'THE CHASTENING of the LORD.' Four Bible Readings given at St. Peter's, Eaton Square. Fcap. 8vo. cloth boards, 1s.
[9th Thousand.

'THE COMMUNION OF SAINTS.' A Help to the Higher Life of Communicants. Fcap. 8vo. cloth boards, 1s.
[22nd Thousand.

THE POWER OF SUFFERING : A Thought for Holy Week. 6d. per Packet of Twelve.

THE POWER OF WEAKNESS: A Thought for Good Friday. Fcap. 8vo. 3d.

THOUGHTS for the DAY of INTERCESSION.
1d.; 6s. per 100. [15th Thousand.

THOUGHTS ON CALVARY. The Substance of Two Good Friday Addresses. Fcap. 8vo. 3d.

TWO ADDRESSES TO COMMUNICANTS. Fcap. 8vo. 6d.

WILKINSON.—*Edited by the Rt. Rev. G. H. WILKINSON, D.D.*

    BREAK UP YOUR FALLOW GROUND. A Help to Self-Examination. Price 3*d*.

    SELF-EXAMINATION QUESTIONS. Founded on the Ten Commandments and the Church Catechism. Price 2*d*.

    SIMPLE PRAYERS FOR DAILY USE FOR YOUNG PERSONS. Price 2*d*.

WILLIAMSON. — THE OUTSTRETCHED HANDS: Good Friday Addresses. By the Rev. ARTHUR WILLIAMSON, M.A., Vicar of St. James's, Norlands. Fcap. 8vo. cloth boards, 1*s*.

WILSON.—SACRA PRIVATA: Private Prayers and Meditations. By the late BISHOP WILSON. 24mo. paper covers, 6*d*.

WOOD.—NUMBER ELEVEN, and Other Stories. By FRANCES H. WOOD. Illustrated. 18mo. cloth boards, 1*s*. 6*d*.

WOODHOUSE.—*Works by the Rev. F. C. WOODHOUSE, M.A., Rector of St. Mary's, Hulme.*

    A MANUAL FOR ADVENT: a few Thoughts for Every Day, and for Christmas and the New Year. Crown 8vo. cloth boards, 3*s*. 6*d*. [Second Edition.

    A MANUAL FOR LENT: Meditations for Every Day, and for the Sundays and Eastertide. Crown 8vo. cloth boards, 3*s*. 6*d*. [Third Edition.

    Besides several notices from the Church papers, the CHRISTIAN WORLD says: ' This is a remarkably good book ; thoughtful, striking, earnest, and deeply interesting. It is far superior to the books of its own class which usually come under our notice. It is not scrappy and incoherent, but really full of power and suggestiveness. The style is always clear and cultured. We believe it to be a book which preachers and intelligent laymen will prize greatly, and read with profit and pleasure. All we can add is—happy the people who have the privilege of listening to such a helpful and thoroughly interesting ministry.'

WYNNE.—SPIRITUAL LIFE IN ITS EARLIER STAGES. Five Lectures for Lent. By Rev. G. R. WYNNE, A.M., Vicar of Holywood. Fcap. 8vo. cloth boards, 1*s*. 6*d*.

THE YOUNG STANDARD-BEARER. An Illustrated Temperance Magazine for Children. Price One Halfpenny Monthly. Volumes, cloth boards, 1*s*. 6*d*. each. 1881-1884. Cloth cases for binding a year's numbers, 6*d*.

Published under the Direction of the Church of England Temperance Society.

CONTENTS.

	Page
Absolution, A Sermon	32
Acts, Lessons on the	1
Adult Classes, Hints to Teachers of	12
Aids to Christian Education	18
" the Study of the Books of Samuel	12
Alice and Her Cross	5
Amethyst	25
Artist	1
Be ye Reconciled to God	32
Bell-Ringers, Hints to	12
Bells, Form of Service for Dedication of	27
Bertha's School-Fellows	2
Bible-Class Manual	31
Boy Hero	14
Brazen Serpent	4
Break up your Fallow Ground	33
Breaking of the Bread	8
Bright Thoughts for the Morning	3
British Burma, Personal Recollections of	29
British Guiana, Legends and Myths of	3
Burial of the Dead, Office of	30
By the Sea of Galilee	18
Can She keep a Secret?	6
Canticles Pointed	14
Central Africa	4
Change-Ringing	29
Change-Ringer's Guide to the Steeples of England	29
Changed Cross	12
" " with Music	12
Chants, A Selection of Single and Double	28
Chastening of the Lord	32
Chatterbox	4
Child-Nature	4
Children Busy, Children Glad	23
Children of the Church (1st and 2nd series)	21
" of the Old Testament	5
Children's Home Hymn-book	5
" School Hymn-book	5
Child's Own Story-book	5
Choirmen, Hints to	12
Chorister's Admission Card	5
Christian Church, History of the	25
" Course	19
" Ministry	10
Church and Dissent	13
" and Nonconformists	13
" Congress Reports	5
" in Relation to Home Reunion	13
Churchman's Almanac for Eight Centuries	31
Churchman's Manual	30
Churchmen and Nonconformists: a Better Understanding between	13
Clockmaker of St. Laurent	6
Cloud and the Star	13
Colonel Rolfe's Story	5
Common-Life Sermons	5
Communicants, Two Addresses to	32
Communion of Saints	32
Confession: A Sermon	32
Confirmation, Office of	30
Consulting the Fates	5

	Page
Convocation Reports	7
Coral Missionary Magazine	7
Counsels of a Godfather	30
" on Prayer	23
Count up the Sunny Days	16
Cuddesdon Manual of Intercession for Missions	7
Daily Family Prayer	14
Daisy Offices and Litany	8
" Service of Song	8
Day of Intercession, Hymns for the	27
" " Suggestions for Observing	16
" " Thoughts for the	32
" " Thoughts for the Sick and Infirm	28
Days that are Past	2
Deare Childe	27
Deb Clinton	6
Devotional Life, Instructions in	32
Dictionary of the English Church	8
Discipline of Temptation	25
Divine Fellowship	8
" Service	31
Doctrine of Incarnation	13
Dogged Jack	22
Dolly's own Story	26
Double Witness of the Church	17
Edith Vernon's Life-Work	9
Eirenicon for the Wesleyans	13
Evening Psalter Pointed for Chanting	16
Extra Services for Use in Church	30
Family Worship for Busy Homes	10
Favourite Story-Book	10
Feild, Life of Bishop	29
First Lady of the Land	10
First Steps to Holy Communion	32
Flying Leaves	10
Following Christ	10
Footprints	10
Forms of Prayer to accompany Sermons	9
Fortune-Teller	5
Four Lads and their Lives	5
Four Little Sixes	16
From Do-nothing Hall to Happy-day House	19
From East to West	27
Golden Steps	11
Good Stories	5
" (New Series)	11
Gospel Missionary	11
Gospels, Story of the	1
Grain of Mustard Seed	11
Great Britain for Little Britons	3
Gregory of the Foretop	5
Happy Sunday Afternoons	11
Hearty Services	20
Helen Morton's Trial, and Timid Lucy	11
Help at Hand	12
Helps by the Way	12

CONTENTS.

	Page
Her Great Ambition	12
High Wages	12
Hindrances and Helps	32
,, to Spiritual Life	23
Hints to Church Workers	12
Holiness to the Lord	13
Holy Baptism, Office of	30
,, Communion	14
,, Communion (Sikes)	25
,, Marriage, Two Addresses on	16
,, Matrimony, Office for Solemnization of	30
Holy Scripture: Temperance and Total Abstinence	14
Holy Week and Easter	32
,, ,, and Easter (Bourdaloue)	3
Home Reunion, Church in Relation to	13
,, ,, A Lecture on	13
,, ,, Sermon on	13
Home Reunion Society's Publications	13
Honor Bright	14
Household Prayers	30
How to Keep Lent	32
,, to Deal with Temptation	32
,, to Pray the Lord's Prayer	16
Index Canonum	11
Inheritance of our Fathers	2
Institution and Induction, Form of	27
Instruction for Junior Classes	1
Is it Peace?	31
Jack Stedman	5
Jem Morrison, and the Village Artist	3
Kalendar Notes	30
King in His Beauty	1
Land of Light	17
Laws of Marriage	11
Lay Missioners, Hints to	12
,, Readers, Hints to	12
Lectures on the Lord's Prayer	23
'Left till called for'	17
Left to Our Father	17
Lent Lectures	32
Lenten and other Sermons	4
,, Sermons, Seven	16
Letter of Commendation	18
Little Fables for Little Folks	18
,, Helps for Daily Toilers	18
,, Lays for Little Lips	18
Lost Piece of Silver	18
Love is of God	27
Lucy Graham	6
Lucy Helmore	30
Mackenzie, Life of the Rev. W. B.	4
Magdalen Psalter	29
Manual for Lent	33
,, for Advent	33
Margaret and Her Friends	21
Marriage Service	18
Martin the Skipper	7
Martin Gay the Singer	5
Minister of Christ in these Latter Days	28
Missionary Conference Reports	19
,, Prayers	20
Mission Field	19
,, Life	19
Missions, Speeches on	31

	Page
Month by Month	20
Mopsa the Fairy	16
More Outlines	23
Morning and Evening Prayer	14
Morning Star	20
Mother's Union	20
,, Warm Shawl	21
My Private Prayer-book	31
N. or M.	20
Necessity of Personal Testimony	31
Nether Stoney	5
Notes on the Church Service	14
Number Eleven	33
Off to California	7
Office for New-Year's Eve	9
Offices for Parochial Use, Five	30
,, of Holy Baptism, Confirmation, Solemnization of Matrimony, and Burial of the Dead	30
Old, Old Story	21
Old Andrew the Peacemaker	6
,, Paths	31
,, Ship	10
Oliver Dale's Decision	21
One of a Covey	21
Only a Girl	16
Ordination, Eve of	4
Our Boys and Girls	21
Our Class Meeting	9
Our Church and Our Country	30
Our Waifs and Strays	21
Outline Lessons for each Sunday	4
Outline Illustrations for Little Ones to Colour. First and Second Series	23
Outline Pictures for Little Painters	19
Outstretched Hands	33
Out of the Way	28
Papal Claims	22
Parables of Our Lord practically set forth	8
,, of the Kingdom	22
Parish Library	6
,, Magazine	6
,, Priest, Private Life, &c.	15
Pastor in Parochiâ	15
Pastoral Work	14
Peace in the Sacraments	13
Peas-Blossom	22
Penitentiary Work	32
Pictures from the Poets	24
Plain Forms of Household Prayer	30
,, Texts for Daily Use	22
,, Words, 1, 2, 3, 4, Series	15
,, as Tracts	15
,, to Children	15
Position and Duty of Non-Abstainers	14
Power of Suffering	32
,, of Weakness	32
Practical Sermons	16
Prayer for the Parish	14
Prayer-book, England's	25
,, its History, Language, and Contents	8
Prayers and Meditations for each Day of the Week	9
Prayers for Children	32
,, ,, on Card	32
,, for Schools	15
Preaching, Lectures on	3

CONTENTS.

Title	Page
Present Christ	22
Primitive Episcopacy, Paper on	13
Prize Bible	23
,, for Boys and Girls	6
Psalter and Canticles, pointed	28
Queen's Shilling	24
Quiet Helper	24
Rainhill Funeral	5
Readings and Devotions for Mothers	24
Relations of Church and State	7
,, of the English Church to Nonconformity	13
Religions of the Africans	25
Resolutions for those recovering from Sickness	15
Revision of the Rubrics	15
Rhoda's Secret	5
Robin and Linnet	24
Rochester Diocesan Directory	24
Rope-Sight	26
Rough Diamond	25
Rubrics, Report on	7
Ruth Halliday	25
Sacra Privata	33
Sacrifice in the Eucharist	20
Sale of Advowsons	7
St. Austin's Court	25
Scripture Readings	15
Self-Examination Questions	33
Selwyn, Life of Bishop	29
Sermons, Doctrinal and Practical	9
,, of the City	23
Service for the Admission of a Chorister	14
Seven Prayers on the Seven Words	31
Shadows of Truth	25
Silvermere Annals	25
Simple Guide to Church Doctrine	20
,, Prayers for Young Persons	33
Sister Louise	26
Sister's Bye-Hours	16
Sixty Sermons	9
,, of Noncomformity	9
Snow Queen	24
Songs and Lyrics for Little Lips	26
Special Services (Holy Communion)	27
,, ,, (Intercession for those at Sea)	27
Spiritual Life in its Earlier Stages	33
Standard Methods in Change-Ringing	26
Stories and Episodes	27
,, they tell Me	21
Students' Gospel Harmony	26
Studies for Stories from Girls' Lives	16
,, in the Church	17
Succession of Episcopal Jurisdiction in England	17
Sufferer's Guide	20
Sunday	28
Sunday-school Teachers, Hints to	12
Tale of the Crusades	7
Teaching of Jesus Christ upon many Subjects of the Christian Verity	30
Temperance Hymns and Songs	28
Thoughts on Calvary	32
Three Cups	29
Topsey Turvy	29
Treble Bob : A Treatise on	26
Trials of Rachel Charlcote	30
True Peoitent	30
,, under Trial	22
Turning-Point of Life	30
Twenty Years in Central Africa	25
Tyrrell, Life of Bishop	2
Under Mother's Wing	30
,, the King's Banner	16
Venables, Life of Bishop	17
Vestry Prayers with a Choir	16
Visiting the Poor and Sick, Hints on	12
Voice of God	7
Voices of Nature	31
Waifs and Strays	21
Watchers on the Longships	7
Watching by the Cross	2
Way of Salvation, Instructions in the	32
Ways of overcoming Temptation	2
We are Seven	2
Week-day Services in Conntry Parishes	16
Week Spent in a Glass Pond	9
Whittingham, Life of Bishop	3
Who is Right?	13
Words of Counsel	18
Worthies of the Church of England	1
Worship in Heaven and on Earth	20
Yonng Men's Bible Classes	18
,, Standard-Bearer	33

WELLS GARDNER, DARTON, & Co., Paternoster Buildings, E.C.

www.ingramcontent.com/pod-product-compliance
Lightning Source LLC
Chambersburg PA
CBHW050201230526
45470CB00001B/183